The
Female
Offender

SECOND EDITION

To Ian Yonge Lind

To my sister, Laura Alba

The Female Offender

Girls, Women, and Crime

SECOND EDITION

Meda Chesney-Lind

University of Hawaii at Manoa

Lisa Pasko

University of Hawaii at Manoa

13644

SAGE Publications
International Educational and Professional Publisher
Thousand Oaks ▪ London ▪ New Delhi

For information:

Sage Publications, Inc.
2455 Teller Road
Thousand Oaks, California 91320
E-mail: order@sagepub.com

Sage Publications Ltd.
6 Bonhill Street
London EC2A 4PU
United Kingdom

Sage Publications India Pvt. Ltd.
B-42, Panchsheel Enclave
Post Box 4109
New Delhi 110 017 India

Printed in the United States of America

Library of Congress Cataloging-in-Publication Data

Chesney-Lind, Meda.
The female offender: Girls, women, and crime / by Meda Chesney-Lind
and Lisa Pasko.— 2nd ed.
    p. cm.
Includes bibliographical references and index.
ISBN 0-7619-2978-9 (Cloth: alk. paper)
ISBN 0-7619-2405-1 (Paper: alk. paper)
    1. Female offenders—United States. 2. Female juvenile delinquents—United
States. 3. Discrimination in criminal justice administration—United States.
I. Pasko, Lisa. II. Title.
HV6046.C54 2003
364.3'74'0973—dc21

                                                          2003006150

This book is printed on acid-free paper.

03   04   05   06   07   10   9   8   7   6   5   4   3   2   1

Acquisitions Editor:	Jerry Westby
Editorial Assistant:	Vonessa Vondera
Production Editor:	Denise Santoyo
Typesetter:	C&M Digitals (P) Ltd.
Indexer:	Teri Greenberg
Cover Designer:	Janet Foulger

CONTENTS

ACKNOWLEDGMENTS

This book, like its first edition, took too long; fortunately, this round there are two of us to share the blame, which is only one of many reasons to collaborate. Also long is the list of folks who have made us think about things, helped us with ideas, and basically kept us honest.

For Meda—I once again have to thank my colleagues in the Women's Studies Program, the Department of Sociology, and at the Social Science Research Institute at the University of Hawaii at Manoa for their support. The freedom to write and think as I do comes from having a great workplace—one that celebrates rather than condemns work on girls and women. Special thanks this round goes to Kathy Berg, Dick Dubanoski, Kathy Ferguson, Konia Freitas, Michael Hamnett, Katherine Irwin, David Johnson, Morris Lai, Nancy Marker, and David Mayeda, and for their encouragement and enthusiasm for my work over the years.

For Lisa—I would like to extend many thanks to friends, family, and colleagues who have continuously given me emotional support and always offered avid interest in the book's completion. To name a few, my parents, Jean and Eugene Pasko, as well as Laura Alba, Christopher Bondy, Marilyn Brown, Janet Davidson, Moira Denike, Terri Hurst, Michael Kohan, JD McWilliams, Don Orban, Andrew Ovenden, Stephen Scheele, Michael Skorupka, Tina Slivka, Stephanie Smith, and most especially, Rick Vonderhaar, for his unending technical and helpful assistance.

Both of us are fortunate in our community. Hawaii is such a rich and wonderful social environment within which to work and live. Close association with the Office of Youth Services and the many social service and public agencies with whom they work has greatly enriched our life and work. Bernie Campbell, David Del Rosario, Rodney Goo, Carl Imakyure, Cheryl Johnson, Dee Dee Letts, Bert Matsuoka, David Nakada, Bob Nakata, Tony Pfaltzgraff,

Suzanne Toguchi, have kept us in touch with the youth of Hawaii and their issues. Jo DesMarets, Marcy Brown, Martha Torney, Marian Tsuji, and Louise Robinson have given us much needed help in understanding the issues for adult women offenders. All of these folks have kept us in the community and closer to the reality we want and need to write about.

No work of this scope, though, could have been considered without an equally rich national and international community of scholars with whom we shared ideas, expressed frustration, and plotted strategies. Many of these folks are scholar/activists so their work is enriched by their commitment to seek not only the truth but also social justice. Meda extends deep thanks here to Christine Alder, Joanne Belknap, Barbara Bloom, Lee Bowker, Mickey Eliason, Kathy Daly, Mona Danner, Walter Dekeseredy, Kim English, Karlene Faith, Laura Fishman, John Hagedorn, Tracy Huling, Ron Huff, Russ Immarigeon, Ken Polk, Dan Macallair, Mike Males, Marc Mauer, Merry Morash, Barbara Owen, Nicky Rafter, Robin Robinson, Andrea Shorter, Brenda Smith, Vinnie Schiraldi, Marty Schwartz, Francine Sherman, and last but certainly not least, Randy Shelden.

Nationally and internationally, practitioner/scholars have insisted that they be listened to as well—to understand how girls and women they work with in their communities live. Here we must thank Ilene Bergsman, Alethea Camp, Ellen Clarke, Elaine DeConstanzo, Sue Davis, Jane Higgins, Elaine Lord, Judy Mayer, Andie Moss, C'ana Petrick, Ann McDiarmid, and Paula Schaefer for keeping this work in touch with their reality. Also, wonderful journalists who care about girls and women have worked with me to publicize their situation while also doing important muckraking work that criminologists should have done and would have in better days. Special thanks here to Gary Craig, Adrian Le Blanc, Kitsie Watterson, Nina Siegal, Elizabeth Mehren, and Marie Ragghianti.

Most important, our heartfelt thanks to the girls and women who found themselves in the criminal justice system for having the courage to speak the truth in the face of extraordinary pain. Many of these girls and women must remain anonymous, but fortunately not all. Thanks, most of all, to Linda Nunes for her friendship after so many years, and for giving the hope that women can make it through such systems and survive with integrity. Thanks also to Dale Gilmartin for her help with the girls' issue and her courage to write about her own experience, and to Michelle Alvey for her strength, courage, and trust. we hope that we've done justice to your insights and your experiences.

Finally, thanks to Jerry Westby for never giving up hope that this book would appear. Thanks also to Denise Santoyo and Vonessa Vondera for the final push over the top, and a huge thanks to our students Christina Woo and Pavela Fiaui for their intellectual and practical support as we tried to pull all this information together.

—Meda Chesney-Lind
Lisa Pasko

⊰ ONE ⊱

INTRODUCTION

—•◦•—

TO MY MOTHER
I hate you for lying
And not even trying
You turned your back on me
But maybe you didn't see
I needed you then
But I don't need you now
I can make it through
Though I don't know how
I guess I don't know what to say
Don't know why I turned away
But I think you know why
I couldn't stay
I'm on my own, starting today

—"Annabelle," age 16, child in need of
supervision, State of Massachusetts[1]

O ver a decade ago, Girls Clubs of America[2] crafted a slogan: "Today's Girls Are Tomorrow's Women" (Girls Inc., 1996). Public relations aside, this was both an important observation and a national call for a clearer focus on girls' lives and girls' problems. Why? Somehow, in all the concern about the situation of women and women's issues during the second wave of feminism, the girls were forgotten.

1

Forgetting about girls is easy for adult women to do. After all, because the problems confronting adult women in the workplace and at home are so staggering (sexual harassment, salary inequity, and domestic violence, to name a few), it is difficult to spare energy to consider how their own childhoods shaped who they became and what choices they ultimately faced. Such lack of concern was particularly clear when reviewing the paucity of information on the lives of economically and politically marginalized girls of today's underclass. In the 1990s and continuing into the 21st century, this lack of information has facilitated a spate of mean-spirited initiatives to control the lives (and especially the sexuality and morality) of young girls, most notably African American and Hispanic girls, who are construed as welfare cheats and violent, drug-addicted gang members (Males, 1994).

Consider the recent and racially different depictions of girls' violence and aggression. As this book will document, when dramatic pictures of girls of color carrying guns, committing violent crimes, and wearing bandannas suddenly appeared in the popular media, there were very few careful studies to refute the vivid images. Additionally and without much critical thought, the current attention on "reviving ophelias" and white girls' deployment of violence also contributes to a characterization of girlhood as riddled with aggression, ferocity, and intragender victimizations. Why? Why this absence of critical thinking about girls, violence, and crime? Criminology has long suffered from what Jessie Bernard has called the "stag effect" (Bernard, 1964, as cited in Smith, 1992, p. 218). Criminology has attracted male (and some female) scholars who want to study and understand outlaw men, hoping perhaps that some of the romance and fascination of this role will rub off. As a result, among the disciplines, criminology is almost quintessentially male.

Feminist criminology challenged the overall masculinist nature of criminology by pointing out two important conclusions. First, women's and girls' crime was virtually overlooked, and female victimization was ignored, minimized, or trivialized. Women and girls existed only in their peripheral existence to the center of study—the male world. Second, whereas historical theorizing in criminology was based on male delinquency and crime, these theories gave little awareness of the importance of gender—the network of behaviors and identities associated with the terms masculinity and femininity—that is socially constructed from relations of dominance, power, and inequality between men and women (Chesney-Lind & Sheldon, 1998; Daly & Chesney-Lind, 1988). Feminist criminology demonstrates how gender matters, not only in terms of one's trajectory into crime but also in terms of how the criminal justice system responds to the offenders under its authority.

Because of the interaction between the stag effect and the relative absence of criminological interest in gender theorizing and girls' issues, this book will show that the study of "delinquency" has long excluded girls' behavior from theory and research. To some extent, adult women offenders have also been ignored because it seemed clear that women committed less criminal behavior. The one exception to this generalization is prostitution, which probably came in for some scrutiny because the study of sexuality became both academically fashionable and easily marketed in the 1970s (Winick & Kinsie, 1971). But aside from a few titillating books on prostitution, the silence about girl and women offenders was more or less absolute. Such a situation, as this book will document, has hidden key information from public view and allowed major shifts in the treatment of women and girls—many on the economic margin—to occur without formidable public discussion and debate. Girls and women do get arrested, tried, and sentenced to prison. In fact, there have been major changes in the way that the United States has handled girls' and women's crime in recent decades that do not necessarily bode well for the girls and women who enter the criminal justice system.

First, for all that they are ignored, girls should no longer be an afterthought in the delinquency equation. In fact, girls remain slightly more than one out of four juvenile arrests in the year 2000 (Federal Bureau of Investigation [FBI], 2002). Despite the fact that girls are one quarter of those brought into the juvenile justice system, they have rarely claimed anywhere near that share of public attention or resources. As an example, a study of delinquency prevention and intervention programs found that only 8.2% served only girls or mostly girls (Lipsey, 1992, p. 106). Although the latter half of the 1990s did see growth in gender-responsive programming as well as national and state conferences gathered to address women offenders' issues, the "get tough on crime" initiatives, particularly for drug offenses, and push for incarceration continue to adversely affect women and girls. In the area of women's crime and punishment, a disturbing reality persists: In 1980, there were about 12,000 women in prison; by 2000, there were over 85,000—more than a sevenfold increase in two decades (Bureau of Justice Statistics, 2001; Maquire & Pastore, 1994, p. 600). Moreover, this imprisonment rate for women continues to grow. In 1990, the incarceration rate for female offenders was 31 out of 100,000 female residents; by 2000, it was 59 out of 100,000

(Bureau of Justice Statistics, 2001). Currently, over 950,000 women are under some kind of criminal justice supervision (National Symposium on Female Offenders, 2000).

But why, you might ask, should "normal" people be concerned about the lives of girls and women who become involved with the criminal justice system and end up in prison? What do these people have to do with normal citizens and their daily lives? There are a couple of ways to answer that question. First, and most important, these girls and women are not that different from normal people. Gibbons (1983), for example, points out that the majority of those in the criminal justice system are actually "ordinary individuals who, for the most part, engage in sporadic and unskilled crimes" (p. 203). As we shall see, this is especially true of the girls and women who are the focus of this book.

The role played by social control agencies—the police, the courts, the prisons—in labeling and shaping the "crime problem" is frequently underestimated. We also often overlook the important role the concept of criminal as "outsider" plays in the maintenance of the existing social order (Becker, 1963; Schur, 1984). Clearly, harsh public punishment of a few "fallen" girls and women as witches and whores has always been integral to enforcement of the boundaries of the "good" girls' and women's place in patriarchal society. Anyone seriously interested in examining women's crime or the subjugation of women, then, must carefully consider the role of the contemporary criminal justice system in the maintenance of modern patriarchy.

Another question to ponder, particularly as we begin to explore the experiences of women and girls in the criminal justice system, is why crime, particularly violent crime, is almost exclusively a male preserve, and why sexual crime and its buffer charges (such as being a juvenile "runaway" or an adult prostitute) are found so exclusively in the female realm.

As this book will demonstrate, the women whose lives are changed by these labels are often the victims of what might be called "multiple marginality" (Vigil, 1995) in that their gender, race, and class has placed them at the economic periphery of society. Understanding the lives and choices of girls and women who find themselves in the criminal justice system also requires a broader understanding of the contexts within which their "criminal" behavior is lodged. There are important links between girls' problems and women's crime—links that are often obscured by approaches that consider "delinquency" and "crime" to be separate and discrete topics.

Recent research, for example, on the backgrounds of adult female offenders reveals the importance of viewing them as people with life histories. A few facts about the lives of adult women in U.S. prisons in 2000 make this point very powerfully. Sixty percent of women under correctional authority reported they were physically or sexually assaulted at some time in their lives, and 69% of these women reported the assault happened before they were 18 (Bureau of Justice Statistics, 1999; National Symposium on Female Offenders, 2000). About a third (32%) of women in prison report being physically or sexually assaulted by a family member, relative, or intimate acquaintance. Other studies have shown that nearly one in five of these women inmates had spent time in the foster care system, that well over half (58%) grew up in homes without both parents present, and the adults abused alcohol and drugs in many of these homes (34%; Snell & Morton, 1994).

Research on the childhoods of adult women offenders reveals how the powerful and serious problems of childhood and adolescent victimization dramatically circumscribe girls' choices. In a number of instances, these same problems set the stage for their entry into youth homelessness, unemployment, drug use, survival sex (and sometimes prostitution), and, ultimately, other more serious criminal acts.

Acts that come to be labeled as delinquent or criminal, as this book will document, are like all other social behaviors—they take place in a world where gender still shapes the lives of young people in very powerful ways. This means that gender matters in girls' lives and that the way gender works varies by the community and the culture into which the girl is born. As we shall see, the choices of women and girls on the margin place them in situations in which they are likely to be swept up into the criminal justice system. Likewise, responses to girls' and women's offending must be placed within the social context of a world that is not fair to women, people of color, or those with low incomes. Because criminology has long been sensitive to the role played by class in crime, it is the introduction of gender and race that now poses new challenges for the field in its attempts to understand women's and men's crime.

The challenge in this book is to keep the criminological focus on the fact that girls and women of different cultures and races live in different situations and, as a result, face different choices than their white counterparts. This also means that in addition to the burdens they shoulder because of their gender (living with sexism), they must shoulder the burdens of racism. Because racism also tends to bring discrimination and poverty, the emphasis on class should not be lost, but it cannot be the only lens through which delinquency and crime are understood (as has historically been the case). Finally, though, the focus on race or culture (difference) should not lead to a "politics of difference" that stresses divisions among women to the exclusion of the commonalities of their gender or class. Ultimately, an overemphasis on difference (or race or culture), although appearing to be race-sensitive, can actually excuse white women's silence about issues that affect their nonwhite counterparts (Barry, 1996).

Whatever the reason, there has certainly been no national outcry about the soaring rates of women's imprisonment. Instead, with little or no public discussion, the correctional establishment has gone about the business of building new women's prisons and filling them. Our hope is that this book will help encourage a critical national discussion of this trend. Specifically, that it will provide the best answers we can find to the important questions that surface in discussions about women's crime and punishment. What led these women into criminal behavior? Are today's women offenders more violent than their counterparts in past decades? Finally, how could such a change in public policy toward women have happened with so little fanfare?

As this book will show, the answer to this last question is not simple but much of it lies in our public discomfort with girl and woman offenders and the secrecy that accompanies modern punishment. Not only do we rarely think of girls and women who get arrested, we also tend to ignore the places where the people we detain and imprison are kept. Prisons are not places most of us look at, and even the citizens of towns that house the largest of these institutions tend to look the other way when they drive by.

Silence also shrouds those held by these institutions. As this book will document, most of the people we arrest, jail, try, and imprison are poor, and because they are poor, they are without a voice. Such silence particularly attends the jailing of women because women are supposed to be "good" and not "bad." Their tragedies, their suffering, and their pain are not news, and most of us want to believe that whatever suffering they do endure is simply their due—that the "system" that processes these people is fair and just. Indeed, if the public gives any thought to crime and punishment, it is generally to complain that the system fails to protect us from crime and is too soft on vicious criminals—whom we imagine to be male, violent, and very much unlike ordinary people.

There is little in our everyday lives to challenge that construction. Every night, we are bombarded with images of egregious and senseless violence and almost every face of those shown committing these senseless acts is young, black, and male. What are we to make of these frightening images of anger and violence out of control in our cities?

The first, extremely important point to make about these constructions is that they are grossly untrue. Crime is down, not up, in American society. A 1995 study by the American Bar Foundation found, for example, that with reference to rates of violent crime, "in no instance is the rate higher than 20 years ago and in most categories it is now substantially lower." Murder rates, example, were higher in 1933 than in 2000 (American Bar Foundation, 1995, p. 4; FBI, 2002).

But how can this be true when the media is so full of violence? The sad fact is that our media, particularly our entertainment media, have discovered that violence, unlike humor or drama, travels well. Movies are, first and foremost, increasingly made for an international market, and violence comes cheaper than other forms of "entertainment." The more television our children watch, the more they (and we) come to believe that the scary, mean world they see in the movies exists outside their doors.

The notion of a mean society is also abetted by local news media that find that "if it bleeds, it leads" journalism takes far less energy than the real work of explaining the complex sources of crime and other social problems. Finally, politicians have discovered the fear of crime and its root cause—unarticulated racism—and have had no qualms about turning this fear/racism to their advantage. Crime has become a code word for race, and being tough on crime has become almost a prerequisite for election, with virtually all politicians taking care not to be "out-crimed" by their opponents.

In their rush to appear tough on crime, our leaders have dramatically increased the penalties for virtually every offense in the books, particularly drug offenses. The prisons then exploded not with new, more vicious criminals, but with the very same petty offenders who used to receive probation for their deeds. The least visible of these offenders are the women we are now jailing.

So, as we enter the 21st century, our nation has the dubious distinction of having the highest incarceration rate in the world, second only to Russia (Sentencing Project, 2002b). This incarceration frenzy particularly devastates African American communities—African Americans constitute 46% of the entire prison population and over one in three Black males under some form of criminal justice supervision on any given day (Mauer, 1999; Sentencing Project, 2002a). Corrections is the fastest-growing item in almost every state budget, robbing money from education, housing, and social services (Donziger, 1996). This mindless spending is fueled not by an increase in crime but by cynical political forces that have exploited the unresolved racial and economic inequities in U.S. society.

How do we begin to challenge the correctional-industrial complex that is rapidly emerging around and feeding on our fear of crime and criminals? First, we must meet the prisoner as a person and listen to her story. As she speaks about her life and her experience of prison, a human face is suddenly super-imposed over the mind-numbing figures.

By focusing attention on the girl and woman offender, this book hopes to fuel a public discussion about the unintended victims of our nation's love affair with incarceration—women of color, whose incarceration rate has doubled in the past decade alone. By focusing specifically on girls and women who commit crimes, perhaps it will be easier to understand what brought them to prison. By understanding their lives, we will see that spending money to end violence against girls and women will go a long way toward reducing women's crime. We will also see that ending the grinding poverty that is destroying some neighborhoods and families, rather than punishing the victims of these forces, will do a great deal to reduce girls' and women's crime. Finally, the book will help to end the invisibility of the girl and woman offender. Our ignorance about their lives and their punishments costs us far more than dollars. In our silence, we begin to deny our own humanity and the humanity of those we imprison.

NOTES

1. From Robinson (1990, p. 181). Used with permission.
2. Girls Clubs of America is now Girls Inc.

TWO

GIRLS' TROUBLES
AND "FEMALE DELINQUENCY"

———⊶•⊷———

Every year, girls account for over a quarter of all arrests of young people in America (FBI, 2002, p. 239). Despite this, the young women who find themselves in the juvenile justice system either by formal arrest or referral are almost completely invisible. Our stereotype of the juvenile delinquent is so indisputably male that the general public, those experts whose careers in criminology have been built studying "delinquency," and those practitioners working with delinquent youth, rarely consider girls and their problems.

The next three chapters argue that this invisibility has worked against young women in several distinct ways. First, as this chapter shows, despite the fact that a considerable number of girls are arrested, explanations for the "causes" of delinquency explicitly or implicitly avoid addressing them. Second, major efforts to reform the way the juvenile justice system handles youth were crafted with no concern for girls and their problems within the system. Finally, although girls are no longer completely forgotten at the academic and policy levels, there still exists a paucity of information on girls' development, survival strategies, and pathways to criminality. This dearth of knowledge means that those who work with girls have little guidance in shaping programs or developing resources that can respond to the problems many girls experience.

TRENDS IN GIRLS' ARRESTS

Why are the girls we arrest unnoticed when, in 2001, they accounted for 29% of all juvenile arrests (FBI, 2002, p. 239)? Much of this has to do with the sorts of delinquent acts girls commit. Though many may not realize it, youth can be taken into custody both for criminal acts and a wide variety of what are often called status offenses.

Status offenses, in contrast to criminal violations, permit the arrest of youth for a wide range of behaviors that violate parental authority: "running away from home"; being "a person in need of supervision," "a minor in need of supervision," "incorrigible," "beyond control," "truant"; or in need of "care and protection." Although not technically crimes, these offenses can result in a youth's arrest and involvement in the criminal justice system. Juvenile delinquents, as a category, include youths arrested for either criminal or noncriminal status offenses. Finally, as this chapter shows, status offenses play a major role in girls' delinquency.

Examining the types of offenses for which youth are actually arrested makes it clear that most youths are arrested for the less serious criminal acts and status offenses. Of the 1.3 million youth arrested in 2001, for example, only 4.4% of these arrests were for such serious violent offenses as murder, rape, robbery, or aggravated assault (FBI, 2002, p. 240). In contrast, over three times that number of juvenile offenders were arrested for a single offense (larceny theft), much of which, particularly for girls, is shoplifting (Shelden & Horvath, 1986).

Table 2.1 presents the 10-year difference in arrests of boys and girls for selected offenses. From this, it can be seen that although less serious offenses dominate both male and female delinquency, trivial offenses, particularly status offenses and larceny theft (shoplifting), are more significant in the case of girls' arrests. For example, status offenses account for 26% of arrests for girls, and larceny accounts for 22%. Comparatively, status offenses and larceny each only account for 15% of boys' arrests.

Looking at Table 2.1, we can see several differences in boys' and girls' arrests over the past ten years. First, larceny and status offenses continue to play a more significant role in girls' official delinquency than in boys'—a trend that has remained consistent over previous decades (see Chesney-Lind & Shelden, 1998). This stability is somewhat surprising because dramatic declines in arrests of youth for these offenses might have been expected as a

Table 2.1   Ten-Year Arrest Rates for Persons Under 18, 1992–2001, by Sex

Offense Charged	Males			Females		
	1992	*2001*	*% change*	*1992*	*2001*	*% change*
Total	945,035	858,416	−9.2	293,892	349,252	+18.8
Index Offenses:						
Murder	1,420	508	−64.2	99	69	−30.3
Rape	3,042	2,312	−24.0	59	30	−45.3
Robbery	17,317	11,695	−32.5	1,552	1,104	−28.9
Aggravated assault	33,230	26,302	−20.8	6,325	7,814	+23.5
Burglary	76,831	44,560	−42.0	8,088	6,896	−22.2
Larceny	187,832	118,832	−36.7	79,478	76,831	−3.3
Motor vehicle theft	40,546	18,650	−54.0	6,371	4,176	−34.5
Arson	5,094	4,650	−8.7	604	628	+4
Total violent crime	55,009	40,817	−25.8	8,035	9,017	+12.2
Total property crime	310,303	186,688	−39.8	94,541	87,931	−7.0
Other Offenses:						
Other assaults	72,894	85,896	+17.8	24,148	40,072	+65.9
Forgery and counterfeiting	2,978	2,143	−28.0	1,647	1,221	+25.9
Fraud	3,439	3,136	−8.8	1,619	1,657	+2.3
Stolen property: buying, receiving, possessing	23,310	12,192	−47.7	2,796	2,082	−25.5
Offenses against family	1,662	3,267	+96.6	780	1,825	+134.0
Prostitution	352	208	−40.9	349	435	+24.6
Embezzlement	228	588	+157.9	189	464	+145.5
Vandalism	74,001	50,336	−32.0	7,067	7,561	+7.0
Weapons (carrying, etc.)	28,063	17,639	−37.1	2,089	1,924	−7.9
Drug abuse violations	40,928	86,065	+110.3	5,683	17,083	+200.6
Gambling	545	251	−53.9	33	21	−36.4
Liquor law violations	46,066	52,341	+13.6	18,111	25,074	+38.4
Driving under the influence	8,263	9,430	+28.6	1,170	2,011	+71.9
Drunkenness	9,331	9,245	−0.9	1,856	2,404	+29.5
Disorderly conduct	49,831	60,295	+21.0	14,554	25,809	+77.3
Vagrancy	1,830	1,102	−39.8	319	257	−19.4
All other offenses	130,985	157,096	+19.9	36,500	55,784	+52.8
Suspicion	1,038	376	−63.8	194	203	+4.6
Curfew/loitering	3,198	39,184	+25.6	12,139	19,020	+56.7
Runaways	44,719	31,528	−29.5	59,581	46,879	−21.3

SOURCE: Federal Bureau of Investigation. (2002). *2001 Uniform Crime Reports* (p. 239). Washington, DC: Author.

result of the passage of the Juvenile Justice and Delinquency Prevention Act in 1974. This act, among other things, encouraged jurisdictions to divert and deinstitutionalize youth charged with noncriminal offenses. Although the number of youth arrested for status offenses did drop considerably in the 1970s (arrests of girls for these offenses fell by 24% and arrests of boys fell by an even greater amount—66%; FBI, 1980, p. 191), this trend was reversed in the 1980s. Between 1985 and 1994, for example, girls' runaway arrests increased by 18%, and arrests of girls for curfew violations increased by 83.1% (FBI, 1995, p. 222). The beginning of the 21st century saw a gradual decrease in runaway arrests for both girls and boys, with a 21.3% decrease in girls' runaway arrests and 29.5% decrease in boys' compared to 1992. Although this drop in runaway offenses occurred, curfew and loitering law violations continued to play a significant role in girls' official delinquency. In 2001, girls' arrests for curfew and loitering law violations were up 56.7% since 1992.

For many years, statistics showing large numbers of girls arrested for status offenses were taken to be representative of the different types of male and female delinquency. However, self-report studies of male and female delinquency (which ask school-age youth if they have committed delinquent acts) do not reflect the dramatic differences in misbehavior found in official statistics. Specifically, it appears that girls charged with these noncriminal status offenses have been, and continue to be, significantly overrepresented in court populations.

Teilmann and Landry (1981) compared girls' number of arrests for runaway and incorrigibility with girls' self-reports of these two activities, and found a 10.4% overrepresentation of girls among those arrested as runaways, and a 30.9% overrepresentation of girls arrested for incorrigibility. From these data, they concluded that girls are "arrested for status offenses at a higher rate than boys, when contrasted to their self-reported delinquency rates" (pp. 74–75). These findings were confirmed in another recent self-report study. Figueira-McDonough (1985) analyzed the delinquent conduct of 2,000 youths and found "no evidence of greater involvement of females in status offenses" (p. 277). Similarly, Canter (1982b) found in a National Youth Survey (NYS) that there was no evidence of greater female involvement, compared to males, in any category of delinquent behavior. Indeed, in this sample, males were significantly more likely than females to report committing status offenses.

At the close of the 20th century, other disconcerting differences arose in arrest trends for boys and girls. Whereas boys' arrests have decreased since

1992, girls' arrests have increased by more than 18%, with the largest increases occurring in simple assault, drug abuse, and liquor law violations. These offense categories now account for 28% of girls' total arrests. In 2001, girls accounted for 18% of overall violent crime committed by juveniles and 16% of drug abuse violations—a respective 6% and 4% increase since 1992. Most troubling for girls is the incidence of arrest for drug abuse offenses—a 200% increase since 1992 (compared to boys' 110% increase). In addition, overall female delinquency court caseloads grew by more than 80% between 1988 and 1997, with girls' drug offense case rates rising 106% (Sickmund, 2000). Although boys still account for the overwhelming majority of violent and drug-related offenses, questions remain: Why this increase for girls? Are girls closing the gender gap in violent behavior and drug and alcohol use?

Looking at risk behavior self-reported by adolescent boys and girls, the answer appears to be "no." If changes in arrests reflected changes in girls' behavior, then we might expect such dramatic variations to be reflected in self-report delinquency data. A comparison of 1993 and 2001 data from the National Youth Risk Behavior Survey contradicts this hypothesis: Girls are not reporting an increased use of violence nor has the gender gap in reported drug and alcohol abuse narrowed. The data presented in Table 2.2 demonstrate that girls (as well as boys) are less likely to employ violent behavior—they report fewer engagements in physical fights, injuries, and weapons in 2001 than in 1993. Additionally, boys and girls are not reporting more episodes of drinking alcohol; both boys and girls report slightly lower lifetime and current use of alcohol in 2001 than in 1993. Although both girls and boys report some increase in marijuana and cocaine use, the gender gap has not narrowed; it has remained basically the same. The percentage difference between boys' and girls' reports of marijuana and cocaine use is similar in 2001 to what it was in 1993 (boys still report using each drug 6% to 8% more frequently than do girls) and cannot explain why drug abuse arrests and caseloads have increased more for girls than for boys.

The issue of girls, their arrest trends, and their involvement in violent, drug, and gang activity is addressed at length in the next chapter. Sufficient to note here is that the gender gap is still obvious when serious crimes of violence are considered and that enforcement practices have probably dramatically narrowed the gap in minor assaults (which can range from schoolyard tussles to relatively serious but not life-threatening assaults). Steffensmeier and Steffensmeier (1980) first noted this trend in the 1970s and commented that "evidence suggests that female arrests for 'other assaults' are relatively

Table 2.2   Percentage of High School Students Who Engaged in Risk Behavior,
Youth Risk Behavior Survey, Years 1993 and 2001

	Males		Females	
	1993	*2001*	*1993*	*2001*
In a physical fight	51.2	43.1	31.7	23.9
Injured in a physical fight	5.2	5.2	2.9	2.7
Carried a weapon	34.3	29.3	9.2	6.2
Lifetime alcohol use	80.9	78.6	80.9	77.9
Current alcohol use	50.1	49.2	45.9	45.0
Lifetime marijuana use	36.8	46.5	28.6	38.4
Current marijuana use	20.6	27.9	14.6	20.0
Lifetime cocaine use	5.5	10.3	4.2	8.4
Current cocaine use	2.3	4.7	1.4	3.7
Offered, sold, given illegal drug(s) on school property	28.5	34.6	19.1	22.7

SOURCE: Youth Risk Behavior Surveillance System (2002). *The 2001 Youth Risk Behavior Survey Summary and Results.* Atlanta, GA: Centers for Disease Control. Available online at: www.cdc.gov.

non-serious in nature and tend to consist of being bystanders or companions to males involved in skirmishes, fights, and so on" (p. 70). These figures strongly suggest that official practices and shifts in these practices over time can dramatically affect the character of girls' official delinquency, as opposed to the actual behaviors girls are committing.

Numbers tell only part of the story. How do we explain these patterns and, in particular, the relative absence of serious property and violent delinquency in girls, as compared to boys? To answer this question, we must turn to the theories of delinquency that have long speculated on causes of delinquent behavior in young people.

BOYS' THEORIES AND GIRLS' LIVES

Although existing delinquency theories were developed to explain the behavior of boys, some contend that these theories can be adapted to explain the behavior of girls as well (Baskin & Sommers, 1993; Canter, 1982a; Figuera-McDonough & Selo, 1980; Rowe, Vazsonyi, & Flannery, 1995; Simons, Miller, & Aigner, 1980; Smith & Paternoster, 1987).

To establish that this notion is problematic, this chapter begins with a brief review of the androcentric bias in the major theories of delinquent behavior, old and new. The need for a model of female delinquency that accounts for rather than ignores gender is then explored by reviewing the available evidence on girls' lives and the relationships between girls' problems and their official delinquency. This model of delinquency draws on the best of the insights from traditional delinquency theory and also incorporates insights from contemporary research on gender, adolescence, and social control. This discussion shows that the extensive focus on disadvantaged males in public settings has meant that girls' victimization and the relationship between that experience and girls' official delinquency has been systematically ignored.

Also missed by all but feminist research has been the central role played by the juvenile justice system in the sexualization of female delinquency and the criminalization of girls' survival strategies. A complete understanding of girls' experiences with the juvenile justice system, it will be suggested, must also include explanations of the official actions of the juvenile justice system. The juvenile justice system should be understood as a major force in the social control of women, because it has historically served to reinforce the obedience of all young women to the demands of familial authority, no matter how abusive or arbitrary.

The earliest academic efforts to explain delinquent behavior were clear and unapologetic efforts to study male delinquents. "The delinquent is a rogue male," declared Albert Cohen in his influential book on gang delinquency written in 1955 (p. 140). More than a decade later, Travis Hirschi (1969), in his equally important book titled *Causes of Delinquency*, relegated women to a footnote that suggested, somewhat apologetically, that "in the analysis that follows, the 'non-Negro' becomes 'white,' and the girls disappear" (pp. 35–36).

One might want to believe that such cavalier androcentrism is no longer found in academic approaches to delinquency. For this reason, two more recent examples are instructive. Tracy, Wolfgang, and Figlio (1985), in *Delinquency Careers in Two Birth Cohorts*, revisit the practice of including only boys in their delinquency cohort:

> The decision was made, therefore, to study delinquency and its absence in a cohort consisting of all boys born in 1945 and residing in Philadelphia from a date no later than their tenth birthday until at least their eighteenth. Girls were excluded, partly because of their low delinquency, and partly because

the presence of the boys in the city at the terminal age mentioned could be conclusively established from the record of their registration for military service. (p. 9)

In essence, they defend the exclusion of girls with arguments that would still be considered valid today, rather than with an expression of regret over the dated and myopic ways of thinking about delinquency.

From the current gang frenzy that has gripped the United States comes yet another example. Martin Sanchez Jankowski's (1991) widely cited *Islands in the Streets* has the following entries in his index under "Women":

– and codes of conduct

– individual violence over

– as "property"

– and urban gangs

One might be tempted to believe that the last entry refers to girl gangs and the emerging literature on girls in gangs (see Campbell, 1984, 1990; Harris, 1988; Quicker, 1983), but the "and" in the sentence is not a mistake. Girls are simply treated as the sexual chattel of male gang members or as an "incentive" for boys to join the gang (because "women look up to gang members"; Jankowski, 1991, p. 53).

Jankowski's work and other current discussions of "gang delinquency" (see Taylor, 1990, 1993) actually revive the sexism that characterized the earliest efforts to understand visible lower-class male delinquency in Chicago over half a century earlier. Early field work on delinquent gangs in Chicago set the stage for decades of delinquency research. Then, too, researchers were only interested in talking to and following the boys. Thrasher (1927) studied over 1,000 juvenile gangs in Chicago. He spends approximately one page out of 600 on the five or six female gangs he encountered in his field observations of juvenile gangs. Thrasher did mention, in passing, two factors he felt accounted for the lower number of girl gangs:

> First, the social patterns for the behavior of girls, powerfully backed by the
> great weight of tradition and custom, are contrary to the gang and its activi-
> ties; and secondly, girls, even in urban disorganized areas, are much more

closely supervised and guarded than boys and are usually well incorporated into the family groups or some other social structure. (p. 228)

During roughly the same period as Thrasher, others in Chicago were crafting another influential approach to delinquency. Beginning in 1929, Clifford R. Shaw and Henry D. McKay used an ecological approach (or "social ecology") to the study of juvenile delinquency. Their impressive work, particularly *Juvenile Delinquency in Urban Areas* (1942), and intensive biographical case studies such as Shaw's *Brothers in Crime* (1938) and *The Jack-roller* (1930), set the stage for much of the subcultural research on delinquency. Within the city of Chicago (and other major cities of the era), these researchers noticed that crime and delinquency *rates* varied by areas of the city (just as today, if one examined a map of a city and placed red dots on where most offenders live and where most crimes occur, they would be clustered in relatively few areas). The researchers found that the highest rates of crime and delinquency were also found in the same areas exhibiting high rates of multiple other social problems, such as single-parent families, unemployment, joblessness, multiple-family dwellings, welfare cases, and low levels of education (Chesney-Lind & Shelden 1998, pp. 81–82). Such a distribution is caused by a breakdown of institutional, community-based controls, which in turn is caused by three general factors: industrialization, urbanization, and immigration. People living within these areas lack a sense of "community" as the local institutions (e.g., schools, families, churches) are not strong enough to better provide nurturing and guidance for the area's children. Within such environments, there develops a subculture of criminal values and traditions that replace conventional values and traditions. Such criminal values and traditions persist over time, regardless of who lives in the area (Chesney-Lind & Shelden, pp. 81–82).

Although this ecological approach to crime—later to be broadened and called "social disorganization"—is an important contribution to sociological theorizing on crime, its genesis focused only on male delinquency. In their ecological work, Shaw and McKay (1942) analyzed the official arrest data on only male delinquents in Chicago and repeatedly referred to these rates as "delinquency rates" (though they occasionally make parenthetical reference to data on female delinquency; p. 356). Similarly, their biographical work traced only male experiences with the law. In Shaw's (1938) *Brothers in Crime*, for example, the delinquent and criminal careers of five brothers are followed for

15 years. In none of these works is any justification given for the equation of male delinquency with delinquency.

Other major theoretical approaches to delinquency also focus on the subculture of lower-class communities as a generating milieu for delinquent behavior. Here again, noted delinquency researchers concentrated either exclusively or nearly exclusively on male lower-class culture. For example, Cohen's (1955) work on the subculture of delinquent gangs, which was written nearly 20 years after Thrasher's, deliberately considers only boys' delinquency. His justification for the exclusion of girls is quite illuminating:

> My skin has nothing of the quality of down or silk, there is nothing limpid or flute-like about my voice, I am a total loss with needle and thread, my posture and carriage are wholly lacking in grace. These imperfections cause me no distress—if anything, they are gratifying—because I conceive myself to be a man and want people to recognize me as a full-fledged, unequivocal representative of my sex. My wife, on the other hand, is not greatly embarrassed by her inability to tinker with or talk about the internal organs of a car, by her modest attainments in arithmetic or by her inability to lift heavy objects. Indeed, I am reliably informed that many women—I do not suggest that my wife is among them—often affect ignorance, frailty and emotional instability because to do otherwise would be out of keeping with a reputation for indubitable femininity. In short, people do not simply want to excel; they want to excel as a man or as a woman. (p. 138)

From this, Cohen concludes that the delinquent response, "however it may be condemned by others on moral grounds, has at least one virtue: it incontestably confirms, in the eyes of all concerned, his essential masculinity" (1955, p. 140). Much the same line of argument appears in Miller's (1958) influential paper on the "focal concerns" of lower-class life with its emphasis on the importance of trouble, toughness, excitement, and so forth. These, the author concludes, predispose poor youth (particularly male youth) to criminal misconduct. However, Cohen's comments are notable for their candor and capture both the allure that male delinquency has had for at least some male theorists and the fact that sexism has rendered the female delinquent irrelevant to their work.

Emphasis on blocked opportunities (sometimes referred to as "strain" or anomie theories) emerged from the work of Robert K. Merton (1938), who stressed the need to consider how some social structures exert a definite pressure on certain people in the society to engage in nonconformist rather than conformist conduct. Building on Durkheim's notion of anomie (the breakdown in moral ties, rules, customs, laws, and the like that occurs in the wake of rapid social change), Merton developed one of the most enduring criminological theories. His theory emphasized that there was a "discrepancy" between culturally defined goals of society and the institutionalized (i.e., legitimate) means to obtain them (Chesney-Lind & Shelden, 1998, pp. 83–84). He referred to the goals as "success" goals, which typically revolve around earning money, achieving status, obtaining material goods, and so on. Unfortunately, Merton declared, the opportunities to reach the goals are not evenly distributed throughout society—not everyone has equal access to legitimate means. In consequence, "strain" or pressure is placed "upon certain persons in the society to engage in nonconformist rather than conformist conduct" (Merton, 1938, p. 672).

The pressure, Merton argued, causes individuals to adapt in one of several ways. They might simply conform and accept both the goals and the means. They might become what Merton termed "innovators"; they accept the goals but seek deviant means to obtain them. They might give up and become what Merton called "retreatists," rejecting both the goals and the means. They might also blindly follow the means while rejecting the goals of becoming what Merton called "ritualists." Finally, they might become "rebels" and try to substitute different definitions of success and invent different means (Chesney-Lind & Shelden, 1998, pp. 83–84). His work influenced delinquency research largely through the efforts of Cloward and Ohlin (1960), who discussed access to "legitimate" and "illegitimate" opportunities for male youth. Female delinquency is problematic for both Merton's theory and its supporters. If women have the same goals as men but are more often blocked from legitimate means because of discrimination, then, following Merton's logic, women experience more strain and should therefore commit more crime. This, however, is not the case. Additionally, women and girls are also ignored by Cloward and Ohlin's work, except in *Delinquency and Opportunity*, where women are blamed for male delinquency. Here, the familiar notion is that boys, "engulfed by a feminine world and uncertain of their own identification tend to 'protest' against femininity" (1960, p. 49).

The work of Edwin Sutherland emphasized the fact that criminal behavior was learned in intimate personal groups. The basic premise to his theory, "differential association," is that criminal behavior, like other forms of human behavior, is learned in association with close, intimate friends. More specifically, the learning of criminal behavior includes learning its techniques and developing the motives, drives, rationalizations, and attitudes pertaining thereto (Chesney-Lind & Shelden, 1998, pp. 86–87). In addition, the motives, drives, rationalizations, and attitudes are learned from definitions of legal codes as favorable or unfavorable toward violation of the law, depending upon the prevailing perspective within one's immediate environment. The key proposition of the differential association theory is that "a person becomes delinquent because of an excess of definitions favorable to violation of law over definitions unfavorable to violation of law" (Sutherland & Cressey, 1978, p. 75). Sutherland's work, particularly the notion of differential association, which also influenced Cloward and Ohlin's work, was also male oriented because much of his work was affected by the case studies he conducted of male criminals. Indeed, in describing his notion of how differential association works, Sutherland (1978) used male examples:

> In an area where the delinquency rate is high a boy who is sociable, gregarious, active, and athletic is very likely to come in contact with the other boys in the neighborhood, learn delinquent behavior from them, and become a gangster. (p. 131)

Finally, the work of Travis Hirschi (1969) on the social bonds that control delinquency ("social control theory") was, as stated earlier, derived from research on male delinquents (though he, at least, studied delinquent behavior as reported by youth themselves rather than studying only those who were arrested). According to Hirschi, those with close bonds to social groups and institutions (e.g., family, school) are the least likely to become delinquent because the bonds help keep people "in check." Four major elements constitute the social bond: (a) *attachment* refers to one's connection (mostly of an emotional kind) to conventional groups, such as one's immediate family, peers, the school, and so on; (b) *commitment* refers to the sort of "investment" one makes in conventional society or, as Toby (1957) once stated, a "stake in conformity" because one stands to lose a great deal (respect from others, the time one has spent preparing for a career, and the like) if one violates the law; (c) *involvement* refers to one's participation in traditional activities, such as going to school, working, and participating in sports, because if one is busy with such activities, presumably there is little time for deviant activities (this is related to the old saying "idleness is the devil's workshop"); and (d) *belief* refers to an acceptance of basic moral values and laws. Briefly, Hirschi found that, with some exceptions, the facts he collected tended to support his theory. Specifically, youths who had the strongest attachments were the most committed, had the strongest belief in conventional moral values and the law, and were the least delinquent. A conspicuous limitation to Hirschi's work lies in his myopic focus on social class, treating race and gender as afterthoughts.

Such a persistent focus on social class and such an absence of interest in gender with regard to delinquency is ironic for two reasons. As even the work of Hirschi demonstrated, and as later studies would validate, a clear relationship between social class position and delinquency is problematic, whereas it is clear that gender has a dramatic and consistent effect on delinquency causation (Hagan, Gillis, & Simpson, 1985).

Efforts to construct a feminist theory of delinquency, as this and the next chapter demonstrates, must first and foremost be sensitive to the life situations of girls. Feminist theories of crime and delinquency must account for the myriad ways that *gender matters*. It is more than mere insertion of gender as a variable or brief commentary of "girls' or women's issues." Feminist criminology requires a critical consideration of the impact of girls' structural positions in a gender and racially stratified society. It requires a deep understanding of the strategies girls use to negotiate and resist patriarchy and how these strategies can determine what crimes girls commit. Failure to consider the existing empirical evidence on girls' lives and behavior can quickly lead to stereotypical thinking and theoretical dead ends. One example of this sort of flawed theory building was the notion that the women's movement was causing an increase in women's crime; this "liberation" or "emancipation" theory was more or less discredited in the 1970s when no empirical support for its major features appeared (especially that women were committing more serious, violent, "masculine" offenses; Gora, 1982; Steffensmeier, 1980). As we shall see in the next chapter, this "liberation" hypothesis has returned with a vengeance in recent discussions of girls in gangs.

A more recent example of this same sort of thinking can be found in the "power-control" model of delinquency (Hagan, Simpson, & Gillis, 1987). In this model, the authors speculate that girls commit less delinquency in part because their behavior is more closely controlled by the patriarchal family. The authors' promising beginning quickly gets bogged down in a very limited definition of patriarchal control (focusing on parental supervision and variations in power within the family). Ultimately, the authors' narrow formulation of patriarchal control results in their arguing that mothers' workforce participation leads to increases in daughters' delinquency because these girls find themselves in more "egalitarian families."

This is a not-too-subtle variation on the "liberation" hypothesis previously mentioned. Now, however, it is the mother's liberation that causes the daughter's crime. Aside from many methodological problems in the study (e.g., the authors assume that most adolescents live in families with both parents, argue that female-headed households are equivalent to upper-status "egalitarian" families where both parents work, and in at least some papers they measure delinquency using a six-item scale that contains no status offense items), there is a more fundamental problem with the hypothesis. There is no evidence to suggest that as women's labor-force participation has increased, girls' delinquency has increased. Indeed, during the past two decades, when both women's labor-force participation and the number of female-headed households soared, aggregate female delinquency, measured by both self-report and official statistics, did not escalate—instead marked *declines* were observed particularly in self-reports of serious female delinquency (Ageton, 1983; Chesney-Lind & Belknap, 2002; Chilton & Datesman, 1987; FBI, 2002; Office of Juvenile Justice and Delinquency Prevention, 1992).

Feminist criminologists have faulted all theoretical schools of delinquency for assuming that male delinquency, even in its most violent forms, was somehow a "normal" response to their situations. Girls who shared the same social and cultural milieu as delinquent boys but who were not delinquents were considered by these theories somehow abnormal or "over-controlled" (Cain, 1989). Essentially, law-abiding behavior on the part of at least some boys and men is taken by these theories as a sign of character, but when women avoid crime and violence, it is an expression of weakness (Naffine, 1987).

None of these traditional theories address the life situations of girls on the economic and political margins because they were not looking at or talking to these girls. So what might be another way to approach the issue of gender and delinquency? First, it is necessary to recognize that girls grow up in a different world than boys (Block, 1984; Orenstein, 1994). Girls are aware very early in life that, although both girls and boys have similar problems, girls "have it heaps worse" (Alder, 1986).

Likewise, girls of color grow up and do gender in contexts very different from those of their white counterparts. Because racism and poverty are often fellow travelers, these girls are forced by their color and their poverty to deal early and often with problems of violence, drugs, and abuse. Their strategies for coping with these problems, often clever, strong, and daring, also tend to place them outside the conventional expectations of white girls (Campbell, 1984; Orenstein, 1994; Robinson, 1990).

The remainder of this chapter and the next deal with two aspects of these gender differences. First, the situation of girls who come into the juvenile justice system charged with status offenses and other trivial offenses is considered. Next, the unique issue of girls' violence and girls' gang membership is explored. These two discussions explicate the unique ways gender, color, and class shape the choices made by girls—choices our society has often criminalized.

CRIMINALIZING GIRLS' SURVIVAL: ABUSE, VICTIMIZATION, AND GIRLS' OFFICIAL DELINQUENCY

Girls and their problems have been ignored for a long time. When gender was considered in criminological theory, it was often a "variable" in the testing of theories devised to explain boys' behavior and delinquency. As a result, few have considered that some, if not many, of the girls who are arrested and referred to court have unique and different problems than boys. Hints of these differences, though, abound.

For example, it has long been known that a major reason for the presence of many girls in the juvenile justice system was because their parents insisted on their arrest. After all, who else would report a youth as having "run away" from home? In the early years, parents were the most significant referral source; in Honolulu, 44% of the girls who appeared in court in 1929–1930 were referred by parents (Chesney-Lind, 1971).

Recent national data, although slightly less explicit, also show that girls are more likely to be referred to court by sources other than law enforcement agencies (such as parents). In 1997, only 15% of youth referred for delinquency offenses, but 53% of youth referred for status offenses, were referred to court by sources other than law enforcement entities. The pattern among youth referred for status offenses, in which girls are overrepresented, is also clear. Over half of the youth referred for running away from home (60% of whom were girls) and 89% of the youth charged with ungovernability (half of whom were girls) were referred by entities outside of law enforcement, compared to only 6% of youth charged with liquor offenses (68% of whom were boys; Poe-Yamagata & Butts, 1996; Pope & Feyerherm, 1982; Puzzanchera et al., 2000). Additionally, girls are more frequently committed for status offenses than are boys: 9% of girls in training schools were committed for status offenses, compared to 1.5% of boys (Poe-Yamagata & Butts, 1996, p. 24).

The fact that parents are often committed to two standards of adolescent behavior is one explanation for these disparities—one that should not be discounted as a major source of tension even in modern families. Despite expectations to the contrary, gender-specific socialization patterns have not changed very much, and this is especially true for parents' relationships with their daughters (Ianni, 1989; Kamler, 1999; Katz, 1979; Orenstein, 1994; Thorne, 1993). Even parents who oppose sexism in general feel "uncomfortable tampering with existing traditions" and "do not want to risk their children becoming misfits" (Katz, 1979, p. 24).

Thorne (1993), in her ethnography of gender in grade school, found that girls were still using "cosmetics, discussions of boyfriends, dressing sexually, and other forms of exaggerated 'teen' femininity to challenge adult, and class and race-based authority in schools" (p. 156). She also found that "the double standard persists, and girls who are overtly sexual run the risk of being labeled sluts" (p. 156).

Contemporary ethnographies of school life echo the validity of these parental perceptions. Orenstein's (1994) observations also point to the durability of the sexual double standard; at the schools she observed, "sex 'ruins' girls; it enhanced boys" (p. 57). Parents, too, according to Thorne (1993), have new reasons to enforce the time-honored sexual double standard. Perhaps correctly concerned about sexual harassment and rape, to say nothing of HIV/AIDS, "parents in gestures that mix protection with punishment, often tighten control of girls when they become adolescents, and sexuality becomes a terrain of struggle between the generations" (Thorne, p. 156). Finally, Thorne notes that as girls use sexuality as a proxy for independence, they sadly and ironically reinforce their status as sexual objects seeking male approval—ultimately ratifying their status as the subordinate sex.

Whatever the reason, parental attempts to adhere to and enforce the sexual double standard will continue to be a source of conflict between them and their daughters. Another important explanation for girls' problems with their parents that has received attention only in more recent years is that of physical and sexual abuse. Looking specifically at the problem of childhood sexual abuse, it is increasingly clear that this form of abuse is a particular problem for girls.

Girls are, for example, much more likely to be the victims of child sexual abuse than are boys. In nearly eight out of ten sexual abuse cases, the victim is female (Flowers, 2001, p. 146). From a review of community studies, Finkelhor and Baron (1986) estimate that roughly 70% of the victims of sexual abuse are female (p. 45). Sexual abuse of girls tends to start earlier than that of boys (Finkelhor & Baron, 1986, p. 48), girls are more likely than boys to be assaulted by a family member (often a stepfather; DeJong, Hervada, & Emmett, 1983; Russell, 1986), and, as a consequence, their abuse tends to last longer than boys' (DeJong et al., 1983). All of these factors cause more severe trauma and dramatic short- and long-term effects in victims (Adams-Tucker, 1982). The effects noted by researchers in this area move from the well-known "fear, anxiety, depression, anger and hostility, and inappropriate sexual behavior" (Browne & Finkelhor, 1986, p. 69) to behaviors that include running away from home, difficulty in school, truancy, drug abuse, pregnancy, and early marriage (Browne & Finkelhor, 1986; Widom & Kuhns, 1996). In addition, girls who have experienced sexual abuse in their families are at greater risk for subsequent sexual abuse later in life (Flowers, 2001).

Herman's (1981) study of incest survivors in therapy found that they were more likely to have run away from home than a matched sample of women whose fathers were "seductive" (33% vs. 5%). Another study of women patients found that 50% of the victims of child sexual abuse, but only 20% of the nonvictim group, left home before the age of 18 (Meiselman, 1978).

National research on the characteristics of girls in the juvenile justice system shows the role played in girls' delinquency by physical and sexual abuse. According to a study of girls in juvenile correctional settings conducted by the American Correctional Association (ACA; 1990), a very large proportion of these girls—about half of whom were of minority backgrounds—had experienced physical abuse (61.2%), and nearly half said that they had experienced this abuse 11 or more times. Many had reported the abuse, but a large number said that either nothing changed (29.9%) or that reporting it just made things worse (25.3%). More than half of these girls (54.3%) had experienced sexual abuse, and for most this was not an isolated incident; a third reported that it happened 3 to 10 times, and 27.4% reported that it happened 11 times or more. Most were 9 years old or younger when the abuse began. Again, although many reported the abuse (68.1%), reporting the abuse tended to cause no change or made things worse (ACA, 1990, pp. 56–58).

Given this history, it should be no surprise that the vast majority ran away from home (80.7%) and that of those who ran, 39% had run away 10 or more times. Over half (53.8%) said they had attempted suicide, and when asked the reason why, said it was because they "felt no one cared" (ACA, 1990, p. 55). Finally, what might be called a survival or coping strategy has been criminalized; girls in correctional establishments reported that their first arrests were typically for running away from home (20.5%) or for larceny theft (25.0%; ACA, 1990, pp. 46–71).

Detailed studies of youth entering the juvenile justice system in Florida have compared the "constellations of problems" of girls and boys (Dembo, Sue, Borden, & Manning, 1995; Dembo, Williams, & Schmeidler, 1993). These researchers found that girls were more likely than boys to have abuse histories and contact with the juvenile justice system for status offenses, whereas boys had higher rates of involvement with various delinquent offenses. Further research on a larger cohort of youth ($N = 2,104$) admitted to an assessment center in Tampa concluded that "girls' problem behavior commonly relates to an abusive and traumatizing home life, whereas boys' law violating behavior reflects their involvement in a delinquent life style" (Dembo et al., 1995, p. 21).

This suggests that many young women are running away from profound sexual victimization at home and, once on the streets, are forced into crime to survive. Girls who are sexually abused are more likely than abused boys to run away from home as a direct result of their sexual victimization. As long-term runaway youth, these girls are more likely to engage in a prostitution lifestyle, which is highly correlated with other future problems and victimizations, such as AIDS, depression, and rape (Flowers, 1987, 2001; Widom & Kuhns, 1996). The average age of entry into prostitution for girls in the United States is 14; this makes sense when considering that the rate of child sexual abuse is highest for girls when they are age 12 to 17 (Flowers, 2001, p. 146; O'Toole & Schiffman, 1997).

Interviews with girls who have run away from home show, very clearly, that they do not have much attachment to their delinquent activities. They are angry about being labeled as delinquent yet engage in illegal acts (Chesney-Lind & Shelden, 1998). A Wisconsin study found that 54% of the girls who ran away found it necessary to steal money, food, and clothing to survive. A few exchanged sexual contact for money, food, or shelter (Phelps, McIntosh, Jesudason, Warner, & Pohlkamp, 1982, p. 67). In their study of runaway youth, McCormack and his colleagues found that sexually abused female runaways were significantly more likely than their nonabused counterparts to engage in delinquent or criminal activities, such as substance abuse, petty theft, and prostitution (McCormack, Janus, & Burgess, 1986, pp. 392–393).

The backgrounds of adult women in prison underscore the important links between women's childhood victimization and their later criminal careers (Snell & Morton, 1994). Women offenders frequently report abuse in their life histories. About half of the women in jail (48%) and 57% of women in state prisons report experiences of sexual and/or physical abuse in their lives (Bureau of Justice Statistics, 1999).

Confirmation of the consequences of childhood sexual and physical abuse on adult female criminal behavior has come from a large quantitative study of 908 individuals with substantiated and validated histories of victimization. Widom (1988) found that abused or neglected girls were twice as likely as a matched group of controls to have an adult crime record (16% vs. 7.5%). The difference was also found among men but was not as dramatic (42% vs. 33%). Men who had been abused were more likely to contribute to the "cycle of violence," having more arrests for violent offenses as adult offenders than the control group. In contrast, when women with abuse backgrounds did become involved with the criminal justice system, their arrests tended to involve property and order offenses (such as disorderly conduct, curfew, and loitering violations; Widom, 1988, p. 17).

Given this information, taking a feminist perspective on the causes of female delinquency seems an appropriate next step. First, like boys, girls are frequently the recipients of violence and sexual abuse. But unlike boys, girls' victimization and their response to that victimization is specifically shaped by their status as young women. Perhaps because of the gender and sexual scripts found in patriarchal families, girls are much more likely than boys to be the victim of family-related sexual abuse. Men, particularly men with traditional attitudes toward women, are likely to consider their daughters or stepdaughters as their sexual property and feel justified in turning their adult sexual power against them (Armstrong, 1994; Finkelhor, 1982). In a society that idealizes inequality in male and female relationships and that venerates youth in women, girls are easily defined as sexually attractive by older men (Bell, 1970). In addition, girls' vulnerability to both physical and sexual abuse is heightened by norms that require that they stay at home where their victimizers have access to them.

Moreover, as we will see in a subsequent chapter, girls' victimizers (usually men) have the ability to invoke official agencies of social control in their efforts to keep young women at home and vulnerable. That is to say, abusers traditionally have been able to use the uncritical commitment of the juvenile justice system to parental authority to force girls to obey them. Girls' complaints about abuse were, until recently, routinely ignored. For this reason, statutes that were originally placed in law to "protect" young people have, in the case of some girls, criminalized their survival strategies. Although they run away from abusive homes, parents can employ agencies to enforce their return. If they persist in their refusal to stay at home, they are incarcerated.

Young women, a large number of whom are on the run from sexual abuse and parental neglect, are forced by the very statutes designed to protect them into the lives of escaped convicts. Unable to enroll in school or take a job to support themselves because they fear detection, young female runaways are forced into the streets. Here, they engage in panhandling, petty theft, and occasional prostitution to survive. Young women in conflict with their parents (often for legitimate reasons) may actually be forced by present laws into petty criminal activity, prostitution, and drug use.

In addition, because young girls (but not necessarily young boys) are defined as sexually desirable—more desirable than their older sisters due to the double standard of aging—their lives on the streets (and their survival strategies) take a unique shape—once again shaped by patriarchal values. It is no accident that girls on the run from abusive homes or on the streets because of profound poverty get involved in criminal activities that exploit their sexual object status. American society has defined youthful, physically perfect women as desirable. This means that girls on the streets, who have little else of value to trade, are encouraged to use this "resource" (Campagna & Poffenberger, 1988). Sexuality becomes their main source of power and sexual services their main commodity. This also means that the criminal subculture views them from this perspective (Miller, 1986).

The previous description is clearly not the "entire" story about female delinquency, but it illustrates a theory that starts with the assumption that experiences that differentiate boys and girls might illuminate perplexing but persistent facts, such as the fact that more female than male status offenders find their way into the juvenile justice system. However, theories that are sensitive to shared aspects of girls' and boys' lives should not be entirely neglected (see Chesney-Lind & Shelden, 1998, for a discussion of how these theories might shed light on female delinquency). Many such theories, though, were crafted without considering the ways gender shapes both boys' and girls' realities, and need to be rethought with gender in mind.

Two additional comments are important here. First, a recent attempt to salvage the theories crafted to explain boys' behavior argues that the theories are correct; girls and boys are raised very differently but if girls were raised like boys and found themselves in the same situations as boys, then they would be as delinquent as boys (Rowe et al., 1995). This seems to be a regression from the insights of Hagan and his associates. Girls and boys inhabit a gendered universe and find themselves in systems (especially families and schools) that regulate their behavior in radically different ways. These differences, in turn, have significant consequences for the lives of girls (and boys). We need to think about these differences and what they mean not only for crime but, more broadly, about the life chances of girls and boys.

In general, the socialization of boys, especially of white privileged boys, prepare them for lives of power (Connell, 1987). The socialization of girls, particularly during adolescence, is very different. Even for girls of privilege, there are dramatic and negative changes in their self-perception that are reflected in lowered achievement in girls in math and science (American Association of University Women, 1992; Orenstein, 1994). Sexual abuse and harassment are just being understood as major, rather than minor, themes in the lives of all girls. The lives of girls of color, as we shall see in the next chapter on girls in gangs, illustrate the additional burdens that these young women face as they attempt to contend with high levels of sexual and physical victimization in the home, and with other forms of neighborhood violence and institutional neglect (Joe & Chesney-Lind, 1995; Orenstein, 1994).

Not surprisingly, work focusing on the lives of girls and women, particularly the data on the extent of girls' and women's victimization, has caused a "backlash" in which some suggest the numbers are inflated and meaningless (Roiphe, 1993; Wolf, 1993). Others have argued that emphasizing victimization constructs girls and women as having no agency (Baskin & Sommers, 1993). Both perspectives (arguably one from the right and another from the left) seek to shift the focus away from the unique experiences of women back to a more familiar and less intellectually and politically threatening terrain of race and class. They also seek to deny to the starkest victims of the sex/gender system the ability to speak about their pain. To say that a person has had a set of experiences (even very violent ones) is not to reduce that person to a mindless pawn of personal history, but rather to fully illuminate the context within which that person moves and makes "choices."

DELINQUENCY THEORY AND GENDER: BEYOND STATUS OFFENSES

Delinquency theory has all but ignored girls and their problems. As a result, few attempts have been made to understand the meaning of girls' arrest patterns and the relationship between these arrests and the very real problems that these arrests mask.

Girls live, play, and go to school in the same neighborhoods as boys, but their lives are dramatically shaped by gender. A glance at the pattern of girls' arrests causes as many questions as answers for those theorizing about girls' defiance. Why is running away such a major part of girls' delinquency and such a minor part of boys' misbehavior? Also interesting is the relative absence of the mainstays of boys' delinquency (such as burglary) and of serious crimes of violence in both self-reported and official female delinquency.

Conventional theories of delinquency seem best situated to explain the relative absence of girls from traditional boys' delinquency. Yet in general, these theories talk about how to learn the skills and attitudes to commit male delinquency and how to get the opportunity to engage in these behaviors. This chapter has described major differences between girls' and boys' official delinquency and has offered a gender-based theory to account for these differences. Having done this, it is now appropriate to turn to a particular form of girls' delinquency—girls' involvement in gangs—to see how applicable traditional and feminist theories are to explain girls' involvement in what many might feel is the most "macho" of boys' delinquent activities.

THREE

GIRLS, GANGS, AND VIOLENCE

Rediscovering the "Liberated Female Crook"

———•———

A lthough arrest statistics still reflect the dominance of status and other trivial offenses in official female delinquency, the early part of the 1990s saw a curious resurgence of interest in girls, often girls of color, engaged in nontraditional, masculine behavior—notably joining gangs, carrying guns, and fighting with other girls. The last half of the 1990s continued this "bad girl" discourse, with an added focus on white girls' aggression as an undiscovered, concealed culture.

The increase in the arrests of girls for "other assaults" added fuel to this fire. Between 1992 and 2001, arrests of girls for this offense increased by 65.9% and now represent 46% of all juvenile "other assaults" arrests (up from 33% in 1992; FBI, 2002). What is going on? Are we seeing a major shift in the behavior of girls and an entry of girls into violent behaviors, including gang violence, that were once the nearly exclusive domain of young boys? As we shall see, this is the conclusion one would draw from the papers and television, but a closer look at the trends presents a more complex view.

THE MEDIA, GIRLS OF COLOR, AND GANGS

Fascination with a "new," violent female offender is not really new. In the 1970s, a notion emerged that the women's movement had "caused" a surge in women's serious crimes, but this discussion focused primarily on an imagined increase in crimes of adult women, usually white women (Chesney-Lind, 1986). The current discussion has settled on girls' commission of violent crimes, often in youth gangs. Indeed, there has been a veritable siege of news stories with essentially the same theme—girls are in gangs and their behavior in these gangs does not fit the traditional stereotype of female delinquency.

On August 2, 1993, for example, in a feature spread on teen violence, *Newsweek* printed a box titled "Girls Will Be Girls" that noted, "Some girls now carry guns. Others hide razor blades in their mouths" (Leslie, Biddle, Rosenberg, & Wayne, 1993, p. 44). Explaining this trend, the article notes that "the plague of teen violence is an equal-opportunity scourge. Crime by girls is on the rise, or so various jurisdictions report" (p. 44). Exactly a year earlier, a short-subject broadcast appeared on a CBS program titled *Street Stories*. "Girls in the Hood," which was a rebroadcast of a story that first appeared in January 1992, opened with this voiceover:

> Some of the politicians like to call this the Year of the Woman. The women you are about to meet probably aren't what they had in mind. These women are active, they're independent, and they're exercising power in a field dominated by men. In January Harold Dowe first took us to the streets of Los Angeles to meet two uncommon women who are members of street gangs. (CBS, 1992)

The beginning of the 21st century did not see an end to the panic of girls in gangs. In June 2001, ABC reported that while nationally gang membership was down in the United States, the Justice Department was alarmed about a growing problem: girl gang membership. ABC maintained that girls are "catching up with boys in this one area," "joining gangs for the same reasons as boys," and doing the same activities as boys: selling drugs and committing murder. The same story that opened with a proclamation of how gang membership was on the decline—as low as 20% in some areas—closed with a fear that the drug-selling, violent gang member—*girl* gang member—is "every-where" (Gibbs, 2001).

These stories are only a few examples of the many media accounts that have appeared since the second wave of the "liberation" hypothesis was launched by journalists. Where did this come from? Perhaps the start was an article titled "You've Come a Long Way, Moll," which appeared in the *Wall Street Journal*, January 25, 1990. This article noted that "between 1978–1988 the number of women arrested for violent crimes went up 41.5%, vs. 23.1% for men. The trend is even starker for teenagers" (Crittenden, 1990, p. A14). The trend was accelerated by the identification of a new, specific version of the liberation hypothesis. "For Gold Earrings and Protection, More Girls Take the Road to Violence," announced the front page of the *New York Times*, in an article that opened as follows:

> For Aleysha J., the road to crime has been paved with huge gold earrings and name-brand clothes. At Aleysha's high school in the Bronx, popularity comes from looking the part. Aleysha's mother has no money to buy her nice things so the diminutive 15 year old steals them, an act that she feels makes her equal parts bad girl and liberated woman. (Lee, 1991, p. A1)

This is followed by the assertion that

> there are more and more girls like Aleysha in troubled neighborhoods in the New York metropolitan areas, people who work with children say. There are more girls in gangs, more girls in the drug trade, more girls carrying guns and knives, more girls in trouble. (Lee, 1991, p. A1)

Whatever the original source, at this point, a phenomenon known as "pack journalism" took over. The *Philadelphia Inquirer*, for example, ran a story subtitled "Troubled Girls, Troubling Violence" on February 23, 1992, that asserted the following:

> Girls are committing more violent crimes than ever before. Girls used to get in trouble like this mostly as accomplices of boys, but that's no longer true. They don't need the boys. And their attitudes toward their crimes are often as hard as the weapons they wield—as shown in this account based on documents and interviews with participants, parents, police and school officials. While boys still account for the vast majority of juvenile crime, girls are starting to catch up. (Santiago, 1992, p. A1)

This particular story featured a single incident in which an African American girl attacked another girl (described as "middle class" and appearing white in the picture that accompanies the story) in a subway. *The Washington*

Post ran a similar story titled "Delinquent Girls Achieving a Violent Equality in D.C." on December 23, 1992 (Lewis, 1992).

In almost all of the stories on this topic, the issue was framed in a similar fashion. Generally, a specific and egregious example of female violence is described. This is then followed by a quick review of the FBI's arrest statistics, showing what appear to be large increases in the number of girls arrested for violent offenses. Finally, there are quotes from "experts," usually police officers, teachers, or other social service workers, but occasionally criminologists, interpreting the events.

Following these print media stories, the number of articles and television shows focused specifically on girls in gangs jumped. Popular talk shows such as *The Oprah Winfrey Show* (November, 1992), *Geraldo* (January, 1993), and *Larry King Live* (March, 1993) devoted programs to the subject, and more recently NBC news broadcast a story on its nightly news that opened with the same link between women's "equality" and girls' participation in gangs:

> Gone are the days when girls were strictly sidekicks for male gang members, around merely to provide sex and money and run guns and drugs. Now girls also do shooting . . . the new members, often as young as twelve, are the most violent. . . . Ironic as it is, just as women are becoming more powerful in business and government, the same thing is happening in gangs. (NBC, 1993)

For many feminist criminologists, this pattern is more than a little familiar. For example, a 1972 *New York Times* article titled "Crime Rate of Women Up Sharply Over Men's" noted that "Women are gaining rapidly in at least one traditional area of male supremacy—crime" (Roberts, 1971, p. 1).

An expanded version of what would come to be known as the "liberation hypothesis" appeared in Adler's (1975b) *Sisters in Crime*, in a chapter titled "Minor Girls and Major Crimes":

> Girls are involved in more drinking, stealing, gang activity, and fighting—behavior in keeping with their adoption of male roles. We also find increases in the total number of female deviances. The departure from the safety of traditional female roles and the testing of uncertain alternative roles coincide with the turmoil of adolescence creating criminogenic risk factors which are bound to create this increase. These considerations help explain the fact that between 1969 and 1972 national arrests for major crimes show a jump for boys of 82 percent—for girls, 306 percent. (p. 95)

The women's crime wave described by Adler (1975b) and, to a lesser extent, by Simon (1975), was definitively refuted by subsequent research (see Gora, 1982; Steffensmeier & Steffensmeier, 1980), but the popularity of this perspective, at least in the public mind, is apparently undiminished. Whether in the 1990s (and continuing into this century) something different was going on, particularly with reference to girls and gangs, remains still to be seen. This chapter now turns to that question.

TRENDS IN GIRLS' VIOLENCE AND AGGRESSION

A review of girls' arrests for violent crime in the past decade (1992–2001; see Table 3.1) initially seems to provide support for the notion that girls are engaged in more violent crime. Although arrests of girls for murder and robbery were both down since 1992, total violent crime rose 12% for girls, especially in the category of aggravated assault, which increased by 23.5%. As previously stated, other assaults increased by 65.9% and offenses against family members rose 134% (FBI, 2002, p. 239). Changes in arrest rates, which adjust for changes in the population of girls in certain time periods, show much the same pattern.

These increases certainly sound notable, but they are considerably less dramatic on closer inspection. First, if we compare both 10-year and 5-year trends in arrests (Table 3.1), we can see that all Index offenses[1] for girls have decreased since 1997 (except arson, which accounts for only 0.2% of female juvenile offenders). Additionally, only small increases in other assaults (7.5%) and drug abuse violations (4.3%) have occurred in the intervening years, and offenses against the family have actually decreased (by 12.9%) since 1997. If girls were becoming increasingly more aggressive and involved in hard drugs, then violent crime and drug abuse violation arrests should have correlated with such a trend and escalated. Indeed, the opposite is true, with offenses only marginally increasing in the past several years or declining altogether.

Second, what is important to note is that although boys' violent crime has gone down in the past 10 and 5 years, boys still constitute 89% of murder and nonnegligent manslaughter arrests, 91% of robbery arrests, and 76% of aggravated assaults. Serious crimes of violence are a very small proportion of all girls' delinquency, and that figure has remained essentially unchanged historically; violent crime is overwhelmingly a male enterprise. Additionally, if we

Table 3.1   Ten-Year and Five-Year Trends in Arrests for Boys and Girls,
1992–2001, and 1997–2001

Offense Charged	Males		Females	
	1992–2001 % Change	1997–2001 % Change	1992–2001 % Change	1997–2001 % Change
Total	−9.2	−10.0	+18.8	−13.3
Index Offenses:				
Murder	−64.2	−50.0	−30.3	−3.5
Rape	−24.0	−13.5	−45.3	−40.9
Robbery	−32.5	−34.0	−28.9	−39.8
Aggravated assault	−20.8	−16.2	+23.5	−1.6
Burglary	−42.0	−31.6	−22.2	−19.5
Larceny	−36.7	−35.0	−3.3	−20.4
Motor vehicle theft	−54.0	−26.4	−34.5	−22.3
Arson	−8.7	−11.0	+4.0	+5.2
Total violent crime	−25.8	−23.0	+12.2	−9.3
Total property crime	−39.8	−32.9	−7.0	−20.3
Other Offenses:				
Other assaults	+17.8	−5.8	+65.9	+7.5
Forgery and counterfeiting	−28.0	−22.9	+25.9	−30.9
Fraud	−8.8	−15.9	+2.3	−22.5
Stolen property: buying, receiving, possessing	−47.7	−38.0	−25.5	−27.3
Offenses against family	+96.6	−9.1	+134.0	−12.9
Prostitution	−40.9	−29.7	+24.6	+13.5
Embezzlement	+157.9	+22.6	+145.5	+24.9
Vandalism	−32.0	−23.2	+7.0	−13.1
Weapons (carrying, etc.)	−37.1	−27.2	−7.9	−13.8
Drug abuse violations	+110.3	−9.0	+200.6	+4.1
Gambling	−53.9		−36.4	
Liquor law violations	+13.6	−12.4	+38.4	−1.3
Driving under the influence	+28.6	+3.2	+71.9	+12.3
Drunkenness	−0.9	−24.4	+29.5	−5.2
Disorderly conduct	+21.0	−25.1	+77.3	−7.3
Vagrancy	−39.8	−28.0	−19.4	−0.3
All other offenses	+19.9	−15.1	+52.8	−4.7
Suspicion	−63.8	−51.4	+4.6	−9.4
Curfew/loitering	+25.6	−29.7	+56.7	−27.0
Runaways	−29.5	−32.5	−21.3	−27.7

SOURCE: Federal Bureau of Investigation. (2002). 2001 *Uniform Crime Reports* (p. 239).
Washington, DC: Author.

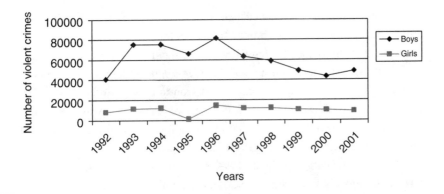

Figure 3.1a   Total Violent Crime* Arrests for Boys and Girls, 1992–2001

SOURCE: Federal Bureau of Investigation. (2002b). *2001 Uniform Crime Reports*. Washington, DC: Author.

Note: * Violent crimes are offenses of murder, forcible rape, robbery, and aggravated assault.

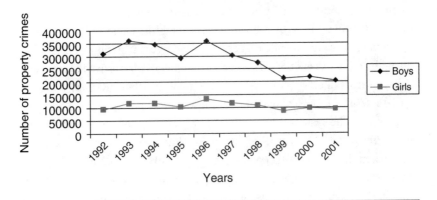

Figure 3.1b   Total Property Crime* Arrests for Boys and Girls, 1992–2001

SOURCE: Federal Bureau of Investigation. (2002b). *2001 Uniform Crime Reports*. Washington, DC: Author.

Note: * Property crimes are offenses of burglary, larceny-theft, motor-vehicle theft, and arson.

compare the overall violent crime and property crime arrests since 1992 for both boys and girls (see Figure 3.1), we find that increases and decreases in girls' arrests more or less parallel fluctuations with boys' arrests (with slightly larger drops in property arrests for boys at the end of the 1990s). Patterns, then, reflect general changes in youth behavior, rather than dramatic changes and shifts in only girls' behavior.

What remains to be explained then is the increase in girls' "other assaults" arrests. Relabeling and "up-criming" behaviors that were once categorized as status offenses into violent offenses cannot be ruled out as a cause for higher assault arrests statistics. A review of the over 2,000 cases of girls referred to Maryland's juvenile justice system for "person-to-person" offenses revealed that almost all of these offenses (97.9%) involved "assault." A further examination of these records revealed that about half were "family centered" and involved such activities as "a girl hitting her mother and her mother subsequently pressing charges" (Mayer, 1994, p. 1). In earlier decades, such behavior would probably have been labeled "incorrigibility" by parents and police. Other mechanisms for relabeling and up-criming status offenses as criminal offenses include police officers advising parents to block the doorways when their children threaten to run away, and then charging the youth with "assault" when they shove past their parents (R. Shelden, personal communication, 1995).

Similar findings can be reached when looking at girls and robbery. A 1998 Honolulu study of robbery arrests for girls suggests that no major shift in the pattern of juvenile robbery occurred over the period 1991 to 1997, which like other jurisdictions had seen the number of girls arrested for robbery increase substantially (Chesney-Lind & Paramore, 1998). Rather, it appears that less serious offenses, particularly those committed by girls, were being swept up into the system. Consistent with this explanation were the following observable patterns: The age of offenders shifted downward, the value of items taken decreased, weapons used were less likely to be lethal in nature, and as a result, fewer injuries to the victims occurred. Most significantly, the proportion of adult victims declined sharply while the number of juvenile victims increased. In short, the study suggested that the problem of female juvenile robbery in the City and County of Honolulu was largely characterized by slightly older youth bullying and "hi-jacking" younger youth for small amounts of cash and, occasionally, jewelry—incidents that had previously been handled informally or formally within the confines of the school system, not law enforcement.

Finally, a detailed study of unpublished FBI data on the characteristics of girls' and boys' homicides between 1984 and 1993 found that girls accounted for "proportionately fewer homicides in 1993 (6%) than in 1984 (14%)" (Loper & Cornell, 1996, p. 324). Detailed comparisons drawn from these "supplemental homicide reports" also indicated that, in comparison to boys' homicides, girls who killed were more likely to use a knife than a gun and to murder someone as a result of conflict rather than in the commission of a crime.

Girls were also more likely than boys to murder family members (32%) and very young victims (24% of their victims were under the age of 3, compared to 1% of boys' victims; Loper & Cornell, 1996, p. 328). When involved in a peer homicide, girls were more likely than boys to have killed "as a result of an interpersonal conflict"; in addition, girls were more likely to kill alone, whereas boys were more likely to kill with an accomplice (p. 328). The authors concluded that "the stereotype of girls becoming gun-toting robbers was not supported. The dramatic increase in gun-related homicides . . . applies to boys but not girls" (p. 16).

Trends in self-report data of youthful involvement in violent offenses also fail to show the dramatic changes. Specifically, a matched sample of "high-risk" youth (ages 13 to 17) surveyed in the 1977 National Youth Study and the 1989 Denver Youth Survey revealed significant decreases in girls' involvement in felony assaults, minor assaults, and hard drugs, and no change in a wide range of other delinquent behaviors—including felony theft, minor theft, and index delinquency (D. Huizinga, personal communication, 1994). As previously mentioned, the 2001 National Youth Risk Behavior Survey expressed similar conclusions—girls report using violence less frequently, report no significant change in alcohol consumption, and report small increases in marijuana and cocaine use that parallel boys' reported increased use as well (Youth Risk Behavior Surveillance System, 2002).

Although many questions can be raised about the actual significance of differences between official and self-report data, careful analyses of these data shed doubt on the media's construction of the hyperviolent girl. However, these data are less helpful in helping us understand girls' involvement with gangs. The reason for this is simple: Changes in official crime statistics and self-report data failed to signal the rise of youth gangs of either gender. As a consequence, it might be more useful to examine other sources of information on gangs and the role of gender in gang membership.

GIRL GANG MEMBERSHIP

Official estimates of the number of youth involved in gangs have increased dramatically over the past decade. Although some accounts of gang activity

Table 3.2   Gang-Related Crime by Type as a Percentage of Total Crime
                  Recorded

Offense Type	Male (%)	Female (%)
Homicide	2.3	4.5
Other violent	48.5	27.3
Property	14.7	42.6
Drug related	10.3	9.1
Vice	2.9	0.0
Other	21.2	16.5
Total	100	100

NOTE: Table derived from data provided by Curry, G.D., Ball, R.A., & Fox, R. J. (1994). *Gang crime and law enforcement recordkeeping* (p. 8). Washington, DC: National Institute of Justice.

report decreased membership since 2000, over 90% of the nation's largest cities still report youth gang problems, up from about half in 1983 (Curry, Fox, Ball, & Stone, 1992; Huff, 2002). Police estimates in large cities now put the number of gangs at 12,538 and the number of gang members at approximately 482,380 (Klein, 2002). But what is the role of gender in gang membership? Let us look more closely at the characteristics of youth labeled by police as gang members.

Curry, Ball, and Fox (1994) reviewed the characteristics of youth identified by police as involved in gangs. Girls account for a very small percentage of these youth (3.6%), though Curry and his associates note that some jurisdictions have "law enforcement policies that officially exclude female gangs members" (Curry et al., p. 8). If only those jurisdictions that include girls and women are examined, the proportion climbs to 5.7%. Table 3.2 displays the offense profiles of male and female gang members in these police databases.

As can be seen, girls are three times more likely than boys to be involved in "property offenses" and about half as likely to be involved in violent offenses. Looking at these statistics differently, girls' involvement in property offenses exceeded 1% of the total number of offenses tracked nationally. Although the table makes much of the role of "homicide" in girls' offenses, only 8 (0.7%) of the 1,072 gang-related homicides in this data set were attributed to girls.

If the jurisdictions that specifically count girls are examined, the numbers change slightly but still show dramatically less involvement of girls in gang-related offenses. In these jurisdictions, girls accounted for only 13.6% of the gang-related property offenses, 12.7% of the drug crimes, and only 3.3%

of the violent crimes. These data do show a higher proportion of girls' involvement in homicides (11.4%), but recall how small the numbers are in this category (Curry et al., 1994, p. 8). Furthermore, remember that detailed research on girls' homicides suggest that their homicides differ significantly from those committed by boys. Girls' homicides are more likely to grow out of an interpersonal dispute with the victim (79%), whereas boys' homicides are more likely to be crime related (57%)—for example, they occur in the commission of another crime, such as robbery (Loper & Cornell, 1996).

A more detailed look at differences between male and female gang members in police databases can be obtained from a study that analyzed files maintained by the Honolulu Police Department (HPD).[2] Examining the characteristics of a sample of youth ($N = 361$) labeled as gang members by the HPD in 1991 (see Chesney-Lind, Rockhill, Marker, & Reyes, 1994, for details of this research), this study specifically looked at the total offense patterns of those labeled as gang members and compared a juvenile subsample of these individuals with nongang delinquents.

In general, Hawaii researchers found patterns consistent with the national data. For example, only 7% of the suspected gang members on Oahu were female and, surprisingly, the vast majority (70%) of these young women were legally adults (the median age was 24.5 for the young women and 21.5 for the young men in the sample).

Almost all the youth identified by police as gang members were drawn from low-income ethnic groups in the islands (Samoan, Native Hawaiian, Filipino), but ethnic differences were also found between male and female gang members. The men were more likely than the women to come exclusively from immigrant groups (Samoan and Filipino); the women were more likely to be Native Hawaiian and Filipino.

Most important, women and girls labeled as gang members committed fewer total numbers of most offenses than men and committed fewer serious offenses. Indeed, the offense profile for the females in the gang sample bears a very close relationship to typical female delinquency. Over a third of the "most serious" arrests of girls (38.1%) were property offenses (larceny theft). This offense category was followed by status offenses (19%) and drug offenses (9.5%). For boys, the most serious offense was likely to be "other assaults" (27%), followed by larceny theft (14%). This profile indicated that although both the boys and girls in this sample of suspected gang members were chronic but not serious offenders, this was particularly true of the girls.

In fact, the pattern of arrests found among these young women imitates national arrest trends of young women. The most common arrest category for girls during this time in the United States was larceny theft (which continued to be the most common arrest category in 2001; FBI, 1995, p. 222; 2002), and the most common arrest category for these girls suspected of gang activity was larceny theft, followed by status offenses. Among boys suspected of gang activity, however, the pattern is somewhat more sobering, with "other assaults" (which often means fighting with other boys) the most serious arrest for the bulk of these young men. In total, serious violent offenses (murder, sexual assault, robbery, and aggravated assault) accounted for 23% of the most serious offenses of boys suspected of gang membership, but for none of the girls' most serious offenses (Chesney-Lind, 1993, p. 336).

Finally, it is important to note that once police identified a youth as a gang member, that youth apparently remained in the database regardless of patterns of desistance. For example, 22% of the sample had not been arrested in 3 years, and there was no gender difference in this pattern.

These patterns prompted further exploration of the degree to which young women labeled by police as "suspected gang members" differed from young women who had been arrested for delinquency. To carry out this exploration, a comparison group was created for those in the Oahu sample who were legally juveniles. Youth suspected of gang membership were matched on ethnicity, age, and gender with youth who were in the juvenile arrest database but who had not been labeled as gang members. A look at offense patterns of this smaller group indicates no major differences between girls suspected of gang membership and their nongang counterparts. The modal most serious offense for gang girls was status offenses, and for nongang girls it was other assaults (Chesney-Lind, 1993, p. 338).

This finding is not totally unexpected. Similar studies, comparing groups in Arizona with Hispanic gangs (Zatz, 1985) and in Las Vegas with African American and Hispanic gangs (Shelden, Snodgrass, & Snodgrass, 1993), although not focusing on gender, found little to differentiate gang members from other "delinquent" or criminal youth. Bowker and Klein (1983), in an examination of data on girls in gangs in Los Angeles in the 1960s, compared the etiology of delinquent behavior of gang girls and their nongang counterparts and asserted the following:

We conclude that the overwhelming impact of racism, sexism, poverty and limited opportunity structures is likely to be so important in determining the gang membership and juvenile delinquency of women and girls in urban ghettos that personality variables, relations with parents and problems associated with heterosexual behavior play a relatively minor role in determining gang membership and juvenile delinquency. (pp. 750–751)

Recent studies of girls in gangs have added to this research. In the late 1990s, studies found gang girls to be engaged in more delinquency than both their female and male nongang counterparts (Miller, 2001). However, gang girls' involvement in the most serious of gang crimes still remained nominal. Deschenes and Esbensen (1999) found that gang membership did increase girls' chances of experiencing violence (both as victims and offenders), but that the frequency of violence used by gang girls was overall relatively low. Girl gang members reported committing robberies or shooting a firearm an average of only once a year and assaulting someone with a weapon twice a year (p. 286).

These quantitative data do not provide support for the rise of a "new" violent female offender and suggest that the hype surrounding the issue has more to do with racism than with crime. Focus on girls in gangs, like its early counterpart, did have one positive effect; it focused much-needed attention on the lives of girls of color. There have been a small but growing number of excellent ethnographic studies of girls in gangs that suggest a much more complex picture wherein some girls solve their problems of gender, race, and class through gang membership. As we review these studies, it will become clear that girls' experiences with gangs cannot simply be characterized as "breaking into" a male world. Girls and women have always been engaged in more violent behavior than the stereotype of women supports; girls have also been in gangs for decades. However, their participation in these gangs, even their violence, is heavily influenced by their gender.

GIRLS AND GANGS: QUALITATIVE STUDIES

Although Curry et al.'s (1994) analysis of official police estimates indicated an extremely small proportion of girls involved in gang activity, other estimates are higher. Miller's (1975) nationwide study of gangs, in the mid-1970s, found the existence of fully independent girl gangs to be quite rare, constituting less than 10% of all gangs, although about half of the male gangs in the New York area had female auxiliary groups. In contrast, Moore's (1991) important ethnographic work on gang activity in Los Angeles' barrios estimated that fully a third of the youth involved in the gangs she studied were female (p. 8).

Given the range of estimates, one might wonder whether girls' involvement with gang life resembles the involvement of girls in other youth subcultures, where they have been described as "present but invisible" (McRobbie & Garber, 1975). Certainly, Moore's (1991) higher estimate indicates that she and her associates saw girls whom others had missed. Indeed, Moore's work is noteworthy in its departure from the androcentric norm in gang research. The longstanding "gendered habits" of researchers have meant that girls' involvement with gangs has been neglected, sexualized, and oversimplified.[3] So, although there have been a growing number of studies investigating the connections among male gangs, violence, and other criminal activities, there has been no parallel development in research on female involvement in gang activity. As with all young women who find their way into the juvenile justice system, girls in gangs have been invisible.

As noted earlier, this pattern of invisibility was undoubtedly set by the initial efforts to understand visible lower-class, male delinquency in Chicago over half a century ago. As an example, Jankowski's (1991) highly regarded *Islands in the Streets* implicitly conceptualizes gangs as a distinctly male phenomenon, and girls are discussed, as noted earlier, in the context of male property:

> In every gang I studied, women were considered a form of property. Interestingly, the women I observed and interviewed told me they felt completely comfortable with certain aspects of this relationship and simply resigned themselves to accepting those aspects they dislike. The one aspect they felt most comfortable with was being treated like servants, charged with the duty of providing men with whatever they wanted. (p. 146)

Taylor's (1993) work, *Girls, Gangs, Women and Drugs*, does focus on girls, but from a distinctly masculine perspective. His work, like Thrasher's and Jankowski's, tends to minimize and distort the motivations and roles of female gang members and is the result of the gender bias of male gang researchers, who describe the female experience from the male gang members' or their own viewpoint (Campbell, 1990). Typically, male gang researchers have characterized female members as maladjusted tomboys or sexual chattel who, in either case, are no more than mere appendages to the male members of the gang.

Taylor's (1993) study provides a veneer of academic support for the media's definition of the girl gang member as a junior version of the liberated female crook of the 1970s. Exactly how many girls and women he interviewed for his book is not clear, but the introduction clearly sets the tone for his work: "We have found that females are just as capable as males of being ruthless in so far as their life opportunities are presented. This study indicates that females have moved beyond the status quo of gender repression" (p. 8). His work stresses the similarities between boys' and girls' involvement in gangs, despite the fact that when the girls and women he interviews speak, it is clear that this view is oversimplified. Listen, for example, to Pat responding to a question about "problems facing girls in gangs":

> If you got a all girls crew, um, they think you're "soft" and in the streets if you soft, it's all over. Fellas think girls is soft, like Rob, he think he got it better in his shit 'cause he's a fella, a man. It's wild, but fellas really hate seeing girls getting off. Now, some fellas respect the power of girls, but most just want us in the sack. (Taylor, 1993, p. 118)

Other studies of female gang delinquency stress that girls have auxiliary roles in boys' gangs (see Bowker, 1978; Brown, 1977; Flowers, 1987; Hanson, 1964; Miller, 1975, 1980; Rice, 1963). Overall, these studies portray girls who are part of gangs as either girlfriends of the male members or "little sister" subgroups of the male gang (Bowker, p. 184; Hanson). Furthermore, they suggest that the role girls play in gangs is "to conceal and carry weapons for the boys, to provide sexual favors, and sometimes to fight against girls who were connected with enemy boys' gangs" (Mann, 1984, p. 45).

Some firsthand accounts of girl gangs, although not completely challenging this image, focus more directly on the race and class issues confronting these girls. Quicker's (1983) study of Chicana gang members in East Los Angeles found that these girls, although still somewhat dependent on their male counterparts, were becoming more independent. These girls identified themselves as "homegirls" and their male counterparts as "homeboys," a common reference to relationships in the barrio. In an obvious reference to "strain theory," Quicker notes that there are few economic opportunities within the barrio to meet the needs of the family unit. As a result, families are disintegrating and cannot provide access to culturally emphasized success goals for young people about to enter adulthood. Not surprisingly, almost all their activities occur within the context of gang life, where they learn how to get along in the world and are insulated within the harsh environment of the barrio (Quicker, 1983).

Moore's (1991) ethnography of two Chicano gangs in East Los Angeles, initiated during the same period as Quicker's (1983), brought the work into the present. Her interviews establish both the multifaceted nature of girls' experiences with gangs in the barrio and the variations in male gang members' perceptions of girls in gangs. Notably, her study establishes that there is no one type of gang girl, with some of the girls in gangs, even in the 1940s, "not tightly bound to boy's cliques" and "much less bound to particular barrios than boys" (p. 27). All the girls in gangs tended to come from a "more troubled background than those of the boys" (p. 30). Significant problems with sexual victimization haunt girls but not boys. Moore documents that the sexual double standard characterized male gang members' and the neighborhood's negative view of girls in gangs (see also Moore & Hagedorn, 1995). Girl gang members were called "tramps" and "no good," despite the girls' vigorous rejection of these labels. Furthermore, some male gang members, even those who had relationships with girl gang members, felt that "square girls were their future" (Moore & Hagedorn, p. 75).

Harris's (1988) study of the Cholas, a Latina gang in the San Fernando Valley, echoes this theme. Although the Cholas resemble male gangs in many respects, the gang challenged girls' traditional destiny within the barrio in two direct ways. First, the girls rejected the traditional image of the Latina woman as "wife and mother," supporting instead a more "macho" homegirl role. Second, the gang supported the girls in their estrangement from organized religion, substituting instead a form of familialism that "provides a strong substitute for weak family and conventional school ties" (p. 172).

The same "macho themes" emerged in a study of the female "age sets" found in a large gang in Phoenix, Arizona (Moore, Vigil, & Levy, 1995). In these groups, fighting is used by girls and boys to achieve status and recognition. Even here, though, the violence is mediated by gender and culture. One girl recounts how she established her reputation by "protecting one of my girls. He [a male acquaintance] was slapping her around and he was hitting her and kicking her, and I went and jumped him and started hitting him" (p. 39). Once respect is earned, these researchers found that girls relied on their reputations and fought less.

Girls in these sets also had to negotiate a Mexican American culture that is "particularly conservative with regard to female sexuality" (Moore et al.,

1995, p. 29). In their neighborhoods and in their relations with the boys in the gang, the persistence of the double standard places the more assertive and sexually active girls in an anomalous position. They must contend with a culture that venerates "pure girls" while also setting the groundwork for the sexual exploitation of girls by gang boys. One of their respondents reports that the boys sometimes try to get girls high and "pull a train" (where a number of boys have sex with one girl), something she clearly objects to, although she admits to having had sex with a boy she didn't like after the male gang members "got me drunk" (p. 32).

Further description of the sexual victimization of girls and women involved in Chicano gangs is supplied by Portillos and Zatz (1995) in their ethnography of Phoenix gangs. They noted that girls can enter gangs either by being "jumped in" or "trained in," the former involving being beaten into the gang and the latter involving having sex with a string of male gang members. Often, those who are "trained in" are later regarded as "loose" and "not really" a gang member. Portillos and Zatz also found extremely high levels of some type of family abuse among the girls they interviewed, which caused them to conclude that "her treatment by male gang members may simply replicate how she is typically treated by males" (p. 24).

Fishman (1995) studied the Vice Queens, an African American female auxiliary gang to a boys' gang, the Vice Kings, that existed in Chicago during the early 1960s. Living in a mostly black community characterized by poverty, unemployment, deterioration, and a high crime rate, the gang of about 30 teenage girls was loosely knit (unlike the male gang) and provided each other with companionship and friends. Failing in school and unable to find work, the girls spent the bulk of their time "hanging out" on the streets with the Vice Kings, which usually included the consumption of alcohol, sexual activities, and occasional delinquency. Most of their delinquency was "traditionally female," such as prostitution, shoplifting, and running away, but some was more serious (e.g., auto theft). They also engaged in fights with other groups of girls, largely to protect their gang's reputation for toughness.

Growing up in rough neighborhoods provided the Vice Queens "with opportunities to learn such traditional male skills as fighting and taking care of themselves on the streets" (Fishman, 1995, p. 87). The girls were expected to learn to defend themselves against "abusive men" and "attacks on their integrity" (p. 87). Their relationship with the Vice Kings was primarily sexual, as sexual partners and mothers of their children, but with no hope of marriage. Fishman perceptively points out that the Vice Queens were

> socialized to be independent, assertive and to take risks with the expecta-
> tions that these are characteristics that they will need to function effectively
> within the black low income community. . . . As a consequence, black girls
> demonstrate, out of necessity, a greater flexibility in roles. (p. 90)

There has been little improvement in the economic situation of the African American community since the 1960s, and today's young women undoubtedly face an even bleaker future than the Vice Queens. In this context, Fishman speculates that "black female gangs today have become more entrenched, more violent, and more oriented to 'male' crime" (p. 91). These changes, she adds, are unrelated to the women's movement but are instead the "forced 'emancipation' which stems from the economic crisis within the black community" (p. 90).

The gender oppression and structural limitations faced by girls in con-temporary poverty-stricken neighborhoods have been largely confirmed by current research by Miller (2001) on girl gang members in Columbus and St. Louis—both relatively new gang cities. With the majority of her gang interviewees African American, Miller found that the segregated and eco-nomically devastated neighborhood environments—which consequently led girls to growing up around crime and gang activity—were influential determi-nants in whether girls joined gangs. Additionally, girl gang members reported that problems within the family, such as drug abuse, violence, and sexual vic-timization, led them to avoid home and join a gang (p. 35). Experiencing gender as both a protective and a risk factor, the girls in Miller's study found gang life empowering as well as victimizing in some ways; they negotiated and strategized gender devaluation within their gangs and social inequality and dangers in their communities. Miller found that girls in gangs not only performed more delinquency and violence, but "it's also the case that gang involvement itself opens up young women to additional victimization risk and exposes them to violence, even when they are not the direct victims, that is sometimes haunting and traumatic in its own right" (p. 151).

Hunt and Joe-Laidler (2001) confirm such findings of victimization and violence in their study of ethnic youth gangs in the San Francisco Bay area. The researchers conclude that "girl gang members experience an extensive amount of violence in their lives whether on the streets, in their family lives, or in their relationships with lovers and boyfriends" (p. 381). Although violence does not consume their everyday lives, girls in gangs do sometimes experience roles of victims (by both men and their own homegirls), perpetrators of, and witnesses to violence. Moreover, these experiences with violence stem from violent-prone situations and life in tension-filled, occasionally hostile neighborhoods, and not the "demonic character" of gang girls themselves (p. 366).

The effects of race and gender, coupled with growing up in economically and socially deprived violent areas, are further illuminated in work by Lauderback, Hansen, and Waldorf (1992) in their study of African American female gangs in San Francisco and by Moore and Hagedorn (1995) in their exploration of ethnic differences between African American and Hispanic female gang members in Milwaukee. Disputing the traditional notions of female gang members as "maladjusted, violent tomboys" and sex objects completely dependent on the favor of male gang members, Lauderback et al. studied an independent girl gang that engaged in crack sales and organized "boosting" to support themselves and their young children (p. 57). All under 25, abandoned by the fathers of their children, abused and controlled by other men, these young women wanted to be "doing something other than selling drugs and to leave the neighborhood" but "many felt that the circumstances which led them to sell drugs were not going to change" (Lauderback et al., p. 69). Enhancing these research findings, Moore and Hagedorn found that when they asked their interviewees if they agreed with the statement, "The way men are today, I'd rather raise my kids myself," 75% of the African American female gang members agreed, compared to only 43% of the Latina gang members. By contrast, 29% of Latinas but none of the African American women agreed that "all a woman needs to straighten out her life is to find a good man" (Moore & Hagedorn, p. 18).

Campbell's work (1984, 1990) on Hispanic gangs in the New York City area further explores the role of the gang for girls in this culture. The girls in her study joined gangs for reasons that are largely explained by their place in a society that has little to offer young women of color (1990, pp. 172–173). First, the possibility of their obtaining a decent career, outside of "domestic servant," was practically nonexistent. Many came from female-headed families subsisting on welfare and most had dropped out of school with no marketable skills. Their aspirations for the future were both sex-typed and unrealistic, with girls wanting to be rock stars or professional models. Second, they found themselves in a highly gendered community where the men in their lives, although not traditional breadwinners, still make many decisions that circumscribe the possibilities open to young women. Third, the responsibilities of young Hispanic mothers further restrict the options available to them. Campbell cites recent data revealing a very bleak future: 94% will have children and 84% will raise their children without a husband. Most will be dependent on some form of welfare (1990, p. 182). Fourth, these young women face a future of isolation as single mothers in the projects. Finally, they share with their male counterparts a future of powerlessness as members of the urban underclass. Their lives, in effect, reflect all the burdens of their triple handicaps of race, class, and gender.

For these girls, Campbell (1990) observes, the gang represents "an idealized collective solution to the bleak future that awaits" them. The girls portray to themselves and the outside world a very idealized and romantic life (p. 173). They develop an exaggerated sense of belonging to the gang. Many were loners prior to joining the gang, only loosely connected to schoolmates and neighborhood peer groups. Yet the gangs' closeness and the excitement of gang life is more fiction than reality. Their daily "street talk" is filled with exaggerated stories of parties, drugs, alcohol, and other varieties of "fun." However, as Campbell notes,

> These events stand as a bulwark against the loneliness and drudgery of their future lives. They also belie the day to day reality of gang life. The lack of recreational opportunities, the long days unfilled by work or school and the absence of money mean that the hours and days are whiled away on street corners. "Doing nothing" means hang out on the stoop; the hours of "bullshit" punctuated by trips to the store to buy one can of beer at a time. When an unexpected windfall arrives, marijuana and rum are purchased in bulk and the partying begins. The next day, life returns to normal. (1990, p. 176)

Joe and Chesney-Lind's (1995) interviews with youth gang members in Hawaii further describe the social role of the gang. Everyday life in marginalized and chaotic neighborhoods sets the stage for group solidarity in two distinct ways. First, the boredom, lack of resources, and high visibility of crime in their neglected communities create the conditions for youth to turn to others who are similarly situated. The group offers a social outlet. At another level, the stress on the family from living in marginalized areas, combined with financial struggles, creates heated tension and, in many cases, violence in the home. Joe and Chesney-Lind found, like Moore, high levels of sexual and physical abuse in the girls' lives: 62% of the girls had been either sexually abused or assaulted. Three fourths of the girls and over half of the boys reported suffering physical abuse.

The group provides both girls and boys with a safe refuge and a surrogate family. Although the theme of marginality cuts across gender and ethnicity, there were critical differences in how girls and boys, and Samoans, Filipinos, and Hawaiians, express and respond to the problems of everyday life. For example, there are differences in boys' and girls' strategies for coping with these pressures—particularly the boredom of poverty. For boys, fighting—even looking for fights—is a major activity within the gang. If anything, the presence of girls around gang members depresses violence. As one 14-year-old Filipino put it, "If we not with the girls, we fighting. If we not fighting, we with the girls" (Joe & Chesney-Lind, 1995, p. 424). Many of the boys' activities involved drinking, cruising, and looking for trouble. This "looking for trouble" also meant being prepared for trouble. Although guns are somewhat available, most of the boys interviewed used bats or their hands to fight, largely but not exclusively because of cultural norms that suggest that fighting with guns is for the weak.

For girls, fighting and violence is a part of their life in the gang but not something they necessarily seek out. Instead, protection from neighborhood and family violence was a consistent and major theme in the girls' interviews. One girl simply stated that she belongs to the gang to provide "some protection from her father" (Joe & Chesney-Lind, 1995, p. 425). Through the group she has learned ways to defend herself physically and emotionally: "He used to beat me up, but now I hit back and he doesn't beat me much now." Another 14-year-old Samoan put it, "You gotta be part of the gang or else you're the one who's gonna get beat up." Although this young woman said that members of her gang had to "have total attitude and can fight," she went on to say, "We want to be a friendly gang. I don't know why people are afraid of us. We're not that violent." Fights do come up in these girls' lives: "We only wen mob this girl 'cause she was getting wise, she was saying 'what, slut' so I wen crack her and all my friends wen jump in" (Joe & Chesney-Lind, pp. 425–426).

Gangs also produce opportunities for involvement in criminal activity, but these are affected by gender as well. Especially for boys from poor families, stealing and small-time drug dealing make up for their lack of money. These activities are not nearly as common among the female respondents. Instead, their problems with the law originate with more traditional forms of female delinquency, such as running away from home. Their families still attempt to hold them to a double standard that results in tensions and disputes with parents that have no parallel among the boys.

LABELING GIRLS VIOLENT?

Historically, those activities that did not fit the official stereotype of "girls' delinquency" have been ignored by authorities (Fishman, 1995; Quicker, 1983; Shacklady-Smith, 1978). Taken together, assessments of gang delinquency in girls, whether quantitative or qualitative, suggest there is little evidence to support the notion of a new, violent female offender. A close reading of ethnographies of gang girls indicates that girls have often been involved in violent behavior as a part of gang life. During earlier periods, however, this occasional violence was ignored by law enforcement officers, who were far more concerned with girls' sexual behavior or morality.

As noted earlier, traditional schools of criminology have assumed that male delinquency, even in its most violent forms, was somehow an understandable if not "normal" response to their situations. This same assumption is not, however, extended to girls who live in violent neighborhoods. If they engage in even minor violence, they are perceived as being more vicious than their male counterparts. In this fashion, the construction of an artificial, passive femininity lays the foundation for the demonization of young girls of color, as shown by the media treatment of girl gang members. Media portrayal of girls and gangs creates a political climate where the victims of racism and sexism can be blamed for their own problems (Chesney-Lind & Hagedorn, 1999). This demonization can then be used as justification for the inattention of these marginalized girls' genuine problems or their horrible treatment in the juvenile justice system.

At best, these ethnographic accounts suggest that girls in gangs are doing far more than seeking "equality" with their male counterparts (Daly & Chesney-Lind, 1988). Girls' involvement in gangs is more than simple rebellion against traditional, white, middle-class notions of girlhood. Girls' gang membership is shaped by the array of economic, educational, familial, and social conditions and constraints that exist in their families and neighborhoods. Indeed, the very structure of the gang and its social life are dependent upon the myriad ways boys and girls manage and construct their gender.

Curry (1995) argues that the discussion of girls' involvement with gangs has tended to go to one extreme or the other. Either girls in gangs are portrayed as victims of injury or they are portrayed as "liberated," degendered gang-bangers. The truth is that both perspectives are partially correct and incomplete without the other. Careful inquiry into the lives of these girls shows the ways in which the gang facilitates survival in their world. In addition, focusing on the social role of the gang in girls' lives illuminates the ways in which girl's and boy's experiences of neighborhood, family, and violence converge and diverge.

GIRLS, GANGS, AND MEDIA HYPE: A FINAL NOTE

A quick comparison of the articles that appeared in each surge of media interest in the "crime wave" committed by girls and women shows many similarities. Most important, those who tout these "crime waves" use a crude form of equity feminism to explain the trends observed and, in the process, contribute to the "backlash" against the women's movement (Faludi, 1991).

There are also crucial differences between the two women's "crime waves." In the stories that announced the first crime wave during the 1970s, the "liberated female crook" was a white political activist, a "terrorist," and a drug-using hippie. For example, one story syndicated by the New York Times service included pictures of both Patty Hearst and Friederike Krabbe (Klemesrud, 1978). Today's demonized woman is a violent African American or Hispanic teenager.

In both instances, there was some small amount of truth in the articles. As this chapter has shown, girls and women have always engaged in more violent behavior than the stereotype of women supports; girls have also been in gangs for decades. The periodic media rediscovery of these facts, then, must be serving other political purposes.

In the past, the goal may have been to discredit young white women and their invisible but central African American counterparts (Barnett, 1993) who were challenging the racism, sexism, and militarism of that day. Today, as the research on girls and gangs has indicated, young minority youth of both genders face a bleak present and a grim future. Today, it is clear that "gang" has become a code word for "race." A review of the media portrayal of girls in gangs suggests that, beyond this, media stories on the youth gang problem can create a political climate in which the victims of racism and sexism can be blamed for their own problems.

In short, this most recent women's "crime wave" appears to be a cultural attempt to reframe the problems of racism and sexism in society. As young women are demonized by the media, their genuine problems can be marginalized and ignored. Indeed, the girls have become the problem. The challenge to those concerned about girls is, then, twofold. First, responsible work on girls in gangs must make the dynamics of this victim blaming clear. Second, it must continue to develop an understanding of girls' gangs that is sensitive to the contexts within which they arise. In an era that is increasingly concerned about the intersections of class, race, and gender, such work seems long overdue.

NOTES

1. Index offenses are defined by the FBI as murder, forcible rape, robbery, burglary, aggravated assault, larceny theft, auto theft, and arson (added in 1979).

2. The Honolulu Police Department, City and County of Honolulu, is located on the island of Oahu, where over three quarters of the state's population resides.

3. For exceptions, see Bowker and Klein (1983), Brown (1977), Campbell (1984, 1990), Fishman (1995), Giordano, Cernkovich, and Pugh (1978), Harris (1988), Moore (1991), Ostner (1986), and Quicker (1983).

FOUR

THE JUVENILE
JUSTICE SYSTEM AND GIRLS

Former Guard Gets One Year in Sexual Abuse of Girl, 14

—The Washington Post, *March 22, 2001*

Abuse Claims Lead to Firings, Inquiry at Jail for Teen Girls

—Chicago Tribune, *May 26, 2001*

State Report Backs up Girl's Fondling Claim

—St. Petersburg Times, *June 22, 2001*

*More Girls Going to Jail—40-Bed Prison Proposed as Colorado's
Judges Hand Down Longer Terms for More Violent Crimes*

—Rocky Mountain News, *January 24, 1999*

Ironically, although the fathers of criminology had little interest in female delinquents during the early part of the 20th century, the same could not be said for the juvenile justice system. Indeed, the early history of the system reveals that concerns about girls' immoral conduct was at the center of what some have called the "childsaving movement" (Platt, 1969) that set up the juvenile justice system.

Half a century later, reforms in the juvenile justice system, particularly in the way that the system handles noncriminal status offenses (running away from home, curfew violation, incorrigibility, etc.) would be crafted. Despite the fact that these offenses, as we saw in Chapter 2, play a major role in girls' experiences with the juvenile justice system, concern about girls would be absent from these discussions as well. Finally, a century after the establishment of the juvenile court, the voices of women's and girls' organizations would force the U.S. Congress to begin asking hard questions about the lives of girls who become labeled delinquents.

This chapter briefly reviews the often invisible experiences of the girls who enter the juvenile justice system (including problems such as those captured in the previous headlines). This chapter also offers some suggestions about how the 21st century might better respond to girls' needs.

"THE BEST PLACE TO CONQUER GIRLS"[1]

The movement to establish separate institutions for youthful offenders was part of the larger Progressive movement that, among other things, was keenly concerned about prostitution and other "social evils" (e.g., white slavery; McDermott & Blackstone, 1994; Rafter, 1990, p. 54; Schlossman & Wallach, 1978). Childsaving was also a celebration of women's domesticity, although, ironically, women were influential in the movement (Platt, 1969; Rafter, 1990).

In a sense, privileged women found, in the moral purity crusades and the establishment of family courts, a safe outlet for their energies. As the legitimate guardians of the moral sphere, women were seen as uniquely suited to patrol the normative boundaries of the social order. Embracing rather than challenging this stereotype, women carved out for themselves a role in the policing of women and girls (Alexander, 1995; Feinman, 1980; Freedman, 1981; Kunzel, 1993; Messerschmidt, 1987). Ultimately, many of the activities of the early childsavers revolved around monitoring the behavior of young girls, particularly immigrant girls, to prevent their straying from the right path.

This state of affairs was the direct consequence of a disturbing coalition between some feminists and the more conservative social purity movement. Concerned about female victimization and suspicious of male (and, to some degree, female) sexuality, notable women leaders, including Susan B. Anthony, found common cause with the social purists around such issues as opposing the regulation of prostitution and raising the age of consent (Messerschmidt, 1987). The consequences of this partnership teach an important lesson to contemporary feminist movements that are, to some extent, faced with the same possible coalitions.

Girls, particularly working-class girls, were the clear losers in this reform effort. Studies of early family court activity reveal that virtually all the girls who appeared in these courts were charged for "immorality" or "waywardness" (Chesney-Lind, 1971; Schlossman & Wallach, 1978; Shelden, 1981). More to the point, the sanctions for such misbehavior were extremely severe. For example, in Chicago (where the first family court was founded), half of the girl delinquents, but only a fifth of the boy delinquents, were sent to reformatories between 1899 and 1909. In Milwaukee, twice as many girls as boys were committed to training schools (Schlossman & Wallach, p. 72), and in Memphis, girls were twice as likely as boys to be committed to training schools (Shelden, p. 70).

In Honolulu during 1929 to 1930, over half of the girls referred to court were charged with "immorality," which meant evidence of sexual intercourse. In addition, another 30% were charged with "waywardness." Evidence of "immorality" was vigorously pursued by both arresting officers and social workers by questioning the girl and, if possible, the boys with whom she was suspected of having sex. Other evidence of "exposure" was provided by the gynecological examinations that were routinely ordered in virtually all girls' cases. Doctors, who understood the purpose of such examinations, would routinely note the condition of the hymen: "admits intercourse, hymen ruptured," "no laceration," and "hymen ruptured" are typical notations on the forms. Girls were also twice as likely as boys to be detained during this period, and they spent, on average, five times as long in detention as their male counterparts. They were also nearly three times more likely than boys to be sentenced to the training school (Chesney-Lind, 1971). Indeed, girls comprised half of those committed to training schools in Honolulu well into the 1950s (Chesney-Lind, 1973).

Not surprisingly, large numbers of reformatories and training schools for girls were established during this period, in addition to places of "rescue and reform." For example, Schlossman and Wallach (1978) note that 23 facilities for girls were opened during the decade 1910 to 1920, in contrast to the period from 1850 to 1910, in which the average was five reformatories per decade

(p. 70). These institutions did much to set the tone of official response to female delinquency. Obsessed with precocious female sexuality, these institutions isolated the girls from all contact with men while housing them in bucolic settings. The intention was to hold the girls until marriageable age and to occupy them in domestic pursuits during their sometimes lengthy incarceration.

So clear was the bias that a few decades into the court's history, astute observers became concerned about the abandonment of minors' rights in the name of treatment, rescue, and protection. One of the most insightful of these critical works, and one that has been unduly neglected, was Paul Tappan's (1947) *Delinquent Girls in Court*. Stating that he was going to look at "what the courts do rather than what they say they do" (p. 2), Tappan evaluated several hundred cases in the Wayward Minor Court in New York City during the late 1930s and early 1940s. These cases caused Tappan to conclude that there were serious problems with a statute that brought young women into court simply for disobedience of parental commands or because they were in "danger of becoming morally depraved" (p. 33). Tappan was particularly concerned that "the need to interpret the 'danger of becoming morally depraved' imposes upon the court a legislative function of a moralizing character" (p. 33). Noting that many young women were being charged simply with sexual activity, he asked, "What is sexual misbehavior—in a legal sense—of the non-prostitute of 16, or 18, or 20 when fornication is no offense under criminal law?" (p. 33).

Tappan (1947) observed that the structure of the Wayward Minor Court "entrusted unlimited discretion to the judge, reformer or clinician and his personal views of expedience" and cautioned that, as a consequence, "the fate of the defendant, the interest of society, the social objectives themselves, must hang by the tenuous thread of the wisdom and personality of the particular administrator" (p. 33). Such an arrangement deeply disturbed Tappan, who noted that "the implications of judicial totalitarianism are written in history" (p. 33).

A more recent historical work on the Los Angeles Juvenile Court during the first half of the 20th century (Odem & Schlossman, 1991) supplies additional evidence of the court's historical preoccupation with girls' sexual morality, along with evidence that the concern clearly colored court activity into the 1950s. Odem and Schlossman reviewed the characteristics of the girls who found their way into this system at two different periods of time during the first half of this century: 1920 and 1950. In 1920, 93% of the girls accused of delinquency were charged with status offenses. Of these, 65% were charged with immoral sexual activity (though the majority of these—56%—had engaged in sex with only one partner, usually a boyfriend). Odem and Schlossman found that 51% of the referrals had originally come from the girl's parents, a situation they explain as caused by working-class parents' fears about their daughter's exposure to the omnipresent temptations to which working-class daughters in particular were exposed to in the modern ecology of urban work and leisure (pp. 197–198). Although working-class girls were encouraged by their families to work (in fact, 52% were currently working or had been working within the past year), their parents were extremely ambivalent about changing community morals and some were not hesitant about involving the court in their arguments with their daughters.

Odem and Schlossman (1991) also found that the Los Angeles Juvenile Court did not shirk from its perceived duty. Seventy-seven percent of the girls were detained prior to their hearing. Both pre- and posthearing detention was common in this court and "clearly linked" to the presence of venereal disease in this population. Data reveal that 35% of all delinquent girls during this period and over half of the alleged sex offenders had gonorrhea, syphilis, or other venereal infections. Odem and Schlossman further note that the presence of disease and the desire to force treatment (which during this period was quite a lengthy and painful proceeding) accounted for the large numbers of girls held in detention centers. Analysis of court actions revealed that although assigning probation was the most common court response (with 61% receiving this outcome), only 27% of girls were released on probation immediately following the hearing. Many girls, it appears, were held in detention centers for weeks or months after their initial hearings.

Girls not placed on probation were often placed in private homes to work as domestics or in a wide range of private institutions, such as the Convent of the Good Shepherd or homes for unmarried mothers. Ultimately, according to this analysis, about 33% of the "problem girls" during this period received a sentence of institutional confinement (Odem & Schlossman, 1991, pp. 198–199).

Between 1920 and 1950, Odem and Schlossman (1991) found that "the make-up of the court's female clientele changed very little" (p. 200). Although the number of black girls brought to court doubled (from 5% to 9%) during this time, the group was still predominantly white (69% compared to 73.5% in 1920), working class, and from disrupted families. Girls were, however, more likely to be in school and less likely to be working in 1950 than in 1920 (p. 200).

Girls referred to court in Los Angeles in 1950 were also overwhelmingly referred for status offenses (78%), though the charges had changed. Now, 31% of the girls were being charged for running away from home, truancy, curfew violations, or "general unruliness at home." Nearly half of the status offenders were charged directly with sexual misconduct, although this was "usually with a single partner; virtually none had engaged in prostitution" (Odem & Schlossman, 1991, p. 200). All these girls were also given physical exams, although the rate of venereal disease had plummeted, with only 4.5% of all girls testing positive. Despite this, the concern for female sexual conduct "remained determinative in shaping social policy" in the 1950s (p. 200).

Referral sources changed within the intervening decades, however, as did sanctions. Parents referred 26% of the girls at mid-decade, school officials about the same percentage in 1950 as 1920 (21% compared to 27%), and police officers referred a greater number in 1950 (54% compared to 29% in 1920). Sanctions shifted slightly, with fewer girls detained prior to hearing in 1950 (56% compared to 77% in 1920), but ultimately the courts ended up placing about the same proportion of girls referred to them in custodial institutions (26% in 1950 compared to 33% in 1920; Odem & Schlossman, 1991).

GIRLS AND JUVENILE JUSTICE REFORM

Problems with the vague nature of status offenses and their sinister meaning for girls continued to haunt the juvenile justice system of the 1960s and 1970s. Status offense categories, students of the court during this period noted, were essentially "buffer charges" for suspected sexuality when applied to girls. Consider the observations of Vedder and Somerville (1970) in their 1960s study of girls in training schools. They found that although girls in their study were incarcerated for the "big five" (running away from home, incorrigibility, sexual offenses, probation violation, and truancy), "the underlying vein in many of these cases is sexual misconduct by the girl delinquent" (p. 147).

Such attitudes were also present in other parts of the world. Naffine (1989) found that, in Australia, official reports noted that, "Most of those charged [with status offenses] were girls who had acquired habits of immorality and freely admitted sexual intercourse with a number of boys" (p. 13). Another study conducted in the early 1970s in a New Jersey training school revealed large numbers of girls incarcerated "for their own protection." When asked about this pattern, one judge explained, "Why most of the girls I commit are for status offenses. I figure if a girl is about to get pregnant, we'll keep her until she's sixteen and then ADC (Aid to Dependent Children) will pick her up" (Rogers, 1972, p. 227).

Andrews and Cohn's (1974) systematic review of the judicial handling of cases of ungovernability in New York in 1972 concluded with the comment that judges were acting "upon personal feelings and predilections in making decisions" (p. 1404). As evidence for this statement, they offer courtroom lectures recorded during the course of their study, such as the following: "She thinks she's a pretty hot number; I'd be worried about leaving my kid with her in a room alone. She needs to get her mind off boys" (p. 1403).

Similar attitudes expressing concern about premature female sexuality and the proper parental response are evident throughout the comments. Another judge remarked that at the age of 14 some girls "get some crazy ideas. They want to fool around with men, and that's sure as hell trouble" (Andrews & Cohn, 1974, p. 1404). Another judge admonished a girl,

> I want you to promise me to obey your mother, to have perfect school atten-
> dance and not miss a day of school, to give up these people who are trying to
> lead you to do wrong, not to hang out in candy stores or tobacco shops or
> street corners where these people are, and to be in when your mother says.
> (p. 1404)

As to where the young woman can go, the judge concluded, in rather telling terms: "I don't want to see you on the streets of this city except with your parents or with your clergyman or to get a doctor. Do you understand?" (p. 1404).

Empirical studies of the processing of girls' and boys' cases that came before the courts between the 1950s and the 1970s clearly documented the effect of these sorts of judicial attitudes. That is, girls charged with status offenses were often more harshly treated than their male or female counterparts charged with crimes (Chesney-Lind, 1973; Cohn, 1970; Datesman & Scarpitti, 1977; Gibbons & Griswold, 1957; Kratcoski, 1974; Mann, 1979; Pope & Feyerherm, 1982; Schlossman & Wallach, 1978; Shelden, 1981). Gibbons and Griswold, for example, found in a study of court dispositions in

Washington state between 1953 and 1955 that although girls were far less likely than boys to be charged with criminal offenses, they were more than twice as likely to be committed to institutions (p. 109). Some years later, a study of a juvenile court in Delaware found that first-time female status offenders were more harshly sanctioned (as measured by institutionalization) than males charged with felonies (Datesman & Scarpitti, p. 70). For repeat status offenders, the pattern became even starker, with females six times more likely than male status offenders to be institutionalized.

This double standard of juvenile justice has also appeared in countries other than the United States. Linda Hancock (1981) found that in Australia, girls (the majority of whom were appearing in court for being uncontrollable and other status offenses) were more likely than boys to receive probation or institutional supervision. In addition, she documented that those girls charged with criminal offenses received lesser penalties than boys and girls brought to court under "protection applications" (p. 8). In England, May (1977) found that girls were less often fined and more often placed on supervision or sent to an institution than boys. In another British study, Smart (1976, p. 134) reported that 64% of girls and 5% of boys were institutionalized for noncriminal offenses. In Portugal, 41% of the girls charged with status offenses in 1984, but only 16.8% of the boys, were placed in institutions (Cain, 1989, p. 222). Likewise, a study of juvenile courts in Madrid revealed that of youth found guilty of status offenses, 22.2% of the girls, but only 6.4% of the boys, were incarcerated (Cain, p. 225).

Careful studies of the juvenile courts well into the second half of the 20th century suggest that judges and other court workers participate rather directly in the judicial enforcement of the sexual double standard. The baldest evidence for this is found in the courts' early years but there is evidence that this pattern continues in many parts of the country. Ironically, these abuses continued, although the same decades ushered in a series of Supreme Court decisions sharply critical of the courts' handling of youthful offenders. Most of this is because the landmark decisions of that era extended to youth charged with crimes—so boys' and not girls' problems were the subject of judicial scrutiny (Chesney-Lind & Shelden, 1998). However, the juvenile justice system's abuse of the status offense category was severely tested, and in some locales eroded, during the 1970s when court critics around the world mounted a major push to "deinstitutionalize and divert" status offenders from formal court jurisdiction.

DEINSTITUTIONALIZATION AND JUDICIAL PATERNALISM: CHALLENGES TO THE DOUBLE STANDARD OF JUVENILE JUSTICE

By the mid-1970s, correctional reformers in many parts of the world became concerned about abuse of the status offense category by juvenile courts. In Victoria, Australia, for example, the 1978 Community Welfare Services Act attempted to remove the more explicitly sexual grounds of some status offenses (notably "exposed to moral danger") and emphasized youths' lack of adequate care and their neglect and abandonment. Limitations were also placed on the courts' authority to find a child "beyond control" of his or her parents (Hancock & Chesney-Lind, 1982, p. 182). South Australia went even further and, in 1979, passed the Children's Protection and Young Offender's Act, which essentially abolished status offenses (Naffine, 1989, p. 10). In Canada, the Province of British Columbia repealed the act permitting incarceration of youth in training schools in 1969 and actively encouraged the "disuse" of that portion of the Federal Juvenile Delinquents Act that dealt with youth found to be "beyond the control of their parents" (Province of British Columbia, 1978, pp. 16–19). Ultimately, Canada replaced the Juvenile Delinquents Act with the Young Offenders Act (1982), which removed status offenders entirely from federal legislation.

In the United States, the Juvenile Justice and Delinquency Prevention (JJDP) Act of 1974 required that states receiving federal delinquency prevention money begin to divert and deinstitutionalize their status offenders. Despite erratic enforcement of this provision and considerable resistance from juvenile court judges, girls were the clear beneficiaries of the reform. Incarceration of young women in training schools and detention centers across the country fell dramatically in the decades following its passage, in distinct contrast to the patterns found early in the century.

National statistics on girls' incarceration reflect both the official enthusiasm for the incarceration of girls during the early part of the twentieth century and the effect of the JJDP Act of 1974. Girls' share of the population of juvenile correctional facilities increased from 1880 (when girls were 19% of the population) to 1923 (when girls were 28%). By 1950, girls had climbed to 34% of the total, and in 1960 they were still 27% of those in correctional facilities. By 1980, this pattern appeared to have reversed, and girls were again 19% of those in correctional facilities (Calahan, 1986, p. 130). In 1991, girls made up

11% of those held in public detention centers and training schools (Moone, 1993a, p. 2).

A separate and more recent analysis shows a leveling-off of this pattern. Poe and Butts (1995) report that girls made up 19% of admissions to detention facilities in the years 1988 through 1992 and 11% of the admissions to long-term facilities (p. 15). More significantly, they report a sharp increase in the percentage of girls' cases involving detentions during roughly the same period (1989–1993). Girls' cases involving detention increased by 23%, compared to an 18% increase in boys' detentions. This is due largely to an increase in the detention of girls for property offenses where, they noted, "growth in female property cases involving detention was more than double the growth among male cases" (p. 12).

These mixed patterns are perhaps a product of the fact that court officials have always been critical of deinstitutionalization (Schwartz, 1989). Not surprisingly, then, although there were great hopes when the Juvenile Justice and Delinquency Prevention Act was passed, a 1978 General Accounting Office (GAO) report concluded that the Law Enforcement Assistance Administration (LEAA), the agency given the task of implementing the legislation, was less than enthusiastic about the deinstitutionalization provisions of the act. Reviewing LEAA's efforts to remove status offenders from secure facilities, the GAO concluded that during certain administrations LEAA had actually "downplayed its importance and to some extent discouraged states from carrying out the Federal requirement" (General Accounting Office, 1978, p. 10).

Just how deep the antideinstitutionalization sentiment was among juvenile justice officials became clear during the House hearings on the extension of the act held in March of 1980. Judge John R. Milligan, representing the National Council of Juvenile and Family Court Judges, argued the following:

> The effect of the Juvenile Justice Act as it now exists is to allow a child ultimately to decide for himself whether he will go to school, whether he will live at home, whether he will continue to run, run, run, away from home, or whether he will even obey orders of your court. (United States House of Representatives, 1980, p. 136)

Ultimately, the judges were successful in narrowing the definition of a status offender in the amended act so that any child who had violated a "valid court order" would no longer be covered under the deinstitutionalization provisions (United States Statutes at Large, 1981). This change, which was never publicly debated in either the House or the Senate, effectively gutted the 1974 JJDP Act by permitting judges to reclassify a status offender who violated a court order as a delinquent. This meant that a young woman who ran away from a court-ordered placement (a halfway house, foster home, etc.) could be relabeled a delinquent and locked up.

Before this change, judges apparently engaged in other, less public, efforts to "circumvent" the deinstitutionalization component of the act. These included "bootstrapping" status offenders into delinquents by issuing criminal contempt citations to elevate status offenders into law violators, referring or committing status offenders to secure mental health facilities, and developing "semi-secure" facilities (Costello & Worthington, 1981, p. 42).

One study that reviewed the effect of these contempt proceedings in Florida (Bishop & Frazier, 1992) found them to work to the disadvantage of female status offenders. This study, which reviewed 162,012 cases referred to juvenile justice intake units during 1985–1987, found only a weak pattern of discrimination against female status offenders, compared to the treatment of male status offenders. However, when they examined the effect of contempt citations, the pattern changed abruptly. Bishop and Frazier found that female offenders referred for contempt were more likely than girls referred for other criminal offenses to be petitioned to court, and substantially more likely than boys referred for contempt to be petitioned to court. Moreover, the girls were far more likely than boys to be sentenced to detention. Specifically, the typical female offender in their study had a 4.3% probability of incarceration, which increased to 29.9% if she was held in contempt. This pattern was not observed for the boys in the study. The authors conclude that

> the traditional double standard is still operative. Clearly neither the cultural changes associated with the feminist movement nor the legal changes illustrated in the JJDP Act's mandate to deinstitutionalize status offenders have brought about equality under the law for young men and women. (Bishop & Frazier, 1992, p. 1186)

Hearings held in conjunction with the most recent reauthorization of the Juvenile Justice and Delinquency Prevention Act, in March 1992, addressed for the first time the "provision of services to girls within the juvenile justice system" (U.S. House of Representatives, 1992, p. 1). At this hearing, the double standard of juvenile justice and the paucity of services for girls were discussed. Representative Matthew Martinez opened the hearing with the following statement:

> In today's hearing we are going to address female delinquency and the provisions of services to girls under this Act. There are many of us that believe that we have not committed enough resources to that particular issue. There are many of us who realize that the problems for young ladies are increasing, ever increasing, in our society and they are becoming more prone to end up in gangs, in crime, and with other problems they have always suffered. (U.S. House of Representatives, 1992, p. 2)

Martinez went on to comment on the high number of girls arrested for status offenses, the high percentage of girls in detention as a result of violation of court orders, and the failure of the system to address girls' needs. He ended with the question, "I wonder why, why are there no other alternatives than youth jail for her?" (U.S. House of Representatives, 1992, p. 2). Testifying at this hearing were also representatives from organizations serving girls, such as Children of the Night, Pace Center for Girls, and Girls Incorporated, in addition to girls active in these programs.

Perhaps as a result of this landmark hearing, the 1992 reauthorization of the JJDP Act of 1974 included specific provisions requiring plans from each state receiving federal funds to include

> an analysis of gender-specific services for the prevention and treatment of juvenile delinquency, including the types of such services available and the need for such services for females and a plan for providing needed gender-specific services for the prevention and treatment of juvenile delinquency. (Public Law 102–586, November 1992)

Additional money was set aside as part of the JJDP Act's challenge grant program for states wanting to develop policies to prohibit gender bias in placement and treatment and to develop programs that ensure girls equal access to services. As a result, 25 states have embarked on such programs—by far the most popular of the 10 possible challenge grant activity areas (Girls Incorporated, 1996, p. 26). The act also called for the GAO to conduct a study of gender bias within state juvenile justice systems, with specific attention to

> the frequency with which females have been detained for status offenses . . . as compared to the frequency with which males have been

detained for such offenses during the 5 year period ending December, 1992; and the appropriateness of the placement and conditions of confinement. (U.S. House of Representatives, 1992, p. 4998)

This mandate will not produce a clear measure of the presence or absence of sexism in the juvenile justice system because it controls for elements of the system that are gendered. Specifically, because girls are overrepresented among those charged with status offenses, "controlling" for status offenses (or more specifically, for the type of status offense) permits discrimination to remain undetected. The mandate does, however, recognize the central role played by status offenses in girls' delinquency (see Chesney-Lind & Shelden, 1992, for a discussion of this problem).

Finally, although not specifically related to gender, the reauthorization of the act moved to make the "bootstrapping" of status offenders into delinquents more difficult. The act specified that youth who were being detained due to a violation of a "valid court order" had to have appeared before a judge and made subject to the order and had to have received, before issuance of the order, "the full due process rights guaranteed to such juvenile by the Constitution of the United States." The act also required that before issuance of the order, "both the behavior of the juvenile being referred and the reasons why the juvenile might have committed the behavior must be assessed." In addition, it must be determined that all dispositions (including treatment), other than placement in a secure detention facility or secure correctional facility, have been exhausted or are clearly inappropriate. Finally, the court has to receive a "written report" stating the results of the review (U.S. House of Representatives, 1992, p. 4983).[2]

Perhaps the most significant changes in the juvenile justice system's handling of girls came indirectly, through the monies Congress made available to states. The 1992 reauthorization, through the "Challenge E" section of this act, set aside funds for states to assess their services to girls. Over 25 states across the United States applied for and received funding to address these goals the most popular of the 10 possible challenge grant activity areas (Girls Incorporated, 1996, p. 26). This state-level funding produced a groundswell of studies, conferences, and programs for girls in states all over the nation (see Chesney-Lind, 2000; Chesney-Lind & Belknap, 2002).

Another outcome of the 1992 reauthorization was funding for a research group to identify "promising practices" and programs regarding delinquent girls. Greene, Peters, and Associates (1998, p. 8) identified five needs that girls require for healthy development: (a) physical safety and healthy development; (b) trust, love, respect, and validation from caring adults; (c) positive female role models; (d) safety to explore their sexual development at their own pace; and (e) feelings of competency, worthiness, and that they "belong." This report also identifies the many hurdles girls face in acquiring these needs, including poverty, family violence, inadequate health care, negative messages about females (particularly their sexuality), and negative community, school, and peer experiences. Stated alternatively, a summary of U.S. and Canadian research identifies "a consistent multi-problem profile of the female young offender," consisting of high rates of both physical and sexual abuse, severe drug addiction, low academic and employment achievement, and chronically dysfunctional and abusive families (Corrado, Odgers, & Cohen, 2000, p. 193). Not surprisingly, as recently as 1998, judges in Ohio reported extremely limited sentencing options for delinquent girls, with almost two-thirds disagreeing with the statement "There are an adequate number of treatment programs for girls." (Less than one third disagreed with this statement about boys; Holsinger, Belknap, & Sutherland, 1999.)

Certainly, these changes mark a distinct departure from previous policies that ignored the situation of girls who found their way into the juvenile justice system. This visibility is clearly needed, because a review of the characteristics of girls in detention centers and training schools still reflects problems with the juvenile justice system's treatment of girls. Specifically, recent research suggests that the new fascination with girls' violence has greatly increased the likelihood that girls will be detained (now for "criminal" offenses). In addition, there is continued evidence that at the level of long-term incarceration, deinstitutionalization has produced a racialized, two-track system of juvenile justice in which white girls are placed in mental hospitals and private facilities, whereas girls of color are institutionalized in public training schools.

RISING DETENTIONS AND RACIALIZED JUSTICE

Although DSO stressed the need to deinstitutionalize status offenders, we have seen that the numbers of girls and boys arrested for these noncriminal offenses continue to remain high, and overall, arrests of girls are increasing dramatically.

Most worrisome are reports of increasing use of detention. National data indicate that between 1989 and 1998, detentions involving girls increased by 56% compared to a 20% increase in boys' detentions (Harms, 2002, p. 1). This same national study attributed the "large increase" in the detention of girls to "the growth in the number of delinquency cases involving females charged with person offenses" (Harms, p. 1). Moreover, these detention rates underline the importance of examining racial differences and discrimination. A study by the American Bar Association and the National Bar Association (2001) reported that half of the girls in secure detention in the United States are African American and 13% are Latina. White girls account for only one-third of the girls in secure detention, although they make up 65% of those in the at-risk population. "Seven of every 10 cases involving white girls are dismissed, compared with 3 of every 10 cases for African American girls" (2001, p. 22).

Girls are also increasingly likely to be officially referred to juvenile courts for these "person" offenses; between 1989 and 1998, the number of girls referred to juvenile courts increased by 83% compared to a 35% increase for males (Stahl, 2001, p. 1). Although girls' referrals outpaced males in all offense categories, the greatest increase was seen in delinquency referrals for "simple assault," which increased by 188% for girls and 108% for males (Stahl, p. 1).

Increases in the detention of girls has also been affected by state initiatives to recriminalize status offenses. As an example, in 1995, the state of Washington passed "Becca's Bill" in the wake of the death of 13-year-old Rebecca Headman, a chronic runaway who was murdered while on the run. Under this legislation, parents can call the police and allege that their daughter has run away. Each time this happens, the girls can be detained in a secure "crisis residential center" for up to 5 days or up to 7 days for contempt if she violated court-ordered conditions. As a result of the passage of this bill, the number of youth placed in detention rose 835% between 1994 and 1997, and estimates are that 60% of the youth taken into custody under "Becca's Bill" are girls "for whom few long-term programs exist" (Sherman, 2000, 2002).

An analysis of official Canadian delinquency data and interviews with Canadian youth probation officers indicate that the rationale used to incarcerate youthful girls is primarily protective, but it is useful to examine this "protection" (Corrado et al., 2000). The scholars in this study hypothesized that although controlling young female offenders' sexuality is part of the patriarchal discrimination, "the sentencing recommendations made by youth justice personnel are primarily based on the desire to protect female youth from high-risk environments and street-entrenched lifestyles" (p. 193). The excusing the practice of detaining in order to protect delinquent girls, then,

> is based partly on the inability of community-based programs to protect certain female youth, the difficulties that these programs have in getting young female offenders to participate in rehabilitation programs, such as drug and alcohol rehabilitation, when they are not incarcerated, and the presence of some, albeit usually inadequate, treatment resources in custodial institutions. As well, cost-effective issues concerning the provision of intervention programs for females engaged in minor offences must also be considered. (Corrado et al., 2000, p. 193)

This study found that when they examined new charges of girls once incarcerated, three-quarters were for "administrative offenses," which are typically probation violations (e.g., failing to attend treatment, obey curfew, and abstain from alcohol).

Thus, although we previously discussed the "bootstrapping" of status offending girls into the system, it is also important to address another current bootstrapping procedure that is often related to status offending and appears to be affecting girls more than boys: serious punishment/sentences for violating court orders. A review of processing girls in the United States is strikingly similar to that of Canadian reports:

> The distinction between delinquency and status offenses has been further muddied by "bootstrapping." A 1980 amendment to the JJDPA allows secure detention for violation of a valid court order, on the basis that such an action (or inaction) constitutes the delinquent offense of contempt. The valid court order holds even when the original offense that brought the young person into the juvenile justice system was a status offense. (Girls Incorporated, 1996, p. 19)

This report recommends that bootstrapping be abolished, given "evidence [that] strongly suggests that bootstrapping results in harsh and inequitable treatment of girls charged with status offenses" (Girls Incorporated, 1996, p. vi). A study by the American Bar Association and the National Bar Association (2001) concluded that girls are not only more likely than boys to be detained

> but to be sent back to detention after release. Although girls' rates of recidivism are lower than those of boys, the use of contempt proceedings and probation and parole violations make it more likely that, without committing a

new crime, girls will return to detention. (p. 20; see Chesney-Lind & Belknap, 2002, for a full discussion of these issues)

San Francisco researchers (Shorter, Schaffner, Schick, & Frappier, 1996) examined the situation of girls in their juvenile justice system and concluded that the girls in their system were "out of sight, out of mind" (p. 1). Specifically, girls would languish in detention centers waiting for placement, while the boys were released or put in placement. As a result, 60% of the girls were detained for more than 7 days, compared to only 6% of the boys (Shorter et al., 1996).

More recently, Acoca and Dedel interviewed 200 girls in county juvenile halls in California. They report "specific forms of abuse" experienced by girls including "consistent use by staff of foul and demeaning language, inappropriate touching, pushing and hitting, isolation, and deprivation of clean clothing" (Acoca, 1999, p. 6). Most disturbing, they report that "some strip searches of girls were conducted in the presence of male officers, underscoring the inherent problem of adult male staff supervising adolescent female detainees" (p. 4).

What of youth in court? According to national estimates provided by the National Center for Juvenile Justice, in 1997 boys constituted 77% of all delinquency referrals to juvenile courts (Puzzanchera et al., 2000). Girls constituted 41% of all petitioned status offenders but only 23% of delinquency referrals (Puzzanchera et al.). Tables 4.1 and 4.2 reveal that although delinquency referrals vary surprisingly little by gender, the status offense categories are heavily gendered. Sixty percent of the youth referred for runaway and 45% of those referred for ungovernability are girls; in contrast, girls are only 32% of those referred for liquor law violations.

National statistics also show that juvenile courts handled well over one and a half million delinquency cases (1,755,000) compared to about 158,500 "petitioned" status offenders (although this figure is described as an estimate). When comparing Table 4.2 to arrests statistics (see Table 2.1 in Chapter 2), these figures indicate that many youth arrested for status offenses are no longer appearing formally before juvenile court judges. In addition, a number of studies now show more evenhanded treatment of boys and girls appearing before court charged with status offenses (see Carter, 1979; Clarke & Koch, 1980; Cohen & Kluegel, 1979; Dungworth, 1977; Johnson & Scheuble, 1991; Teilmann & Landry, 1981).

The picture, however, is not entirely rosy. The number of girls arrested for these and other offenses, for example, continues to climb. Between 1992 and

Table 4.1   National Delinquency Referrals by Sex, 1997

Offense Type	Male (%)	Female (%)	% Female of Total Male/Female Referrals
Person	21	25	26
Property	48	49	24
Drug	12	7	15
Public order	19	20	24
Delinquency	100	100	23
Total	**1,342,900**	**412,100**	

SOURCE: Puzzanchera, C., Stahl, A. L., Finnegan, T. A., Snyder, H. N., Poole, R. S., & Tierney, N. (2000). *Juvenile Court Statistics* 1997. Washington, DC: National Center for Juvenile Justice.

Table 4.2   National Estimates of Petitioned Status Offenders by Sex, 1997

Offense	Male (%)	Female (%)	% Female of Total Male/Female Referrals
Runaway	10	22	60
Truancy	54	54	47
Ungovernability	13	14	45
Liquor	30	20	32
Miscellaneous	24	15	31
Total	**100**	**100**	
	(92,7000)	**(65,8000)**	

SOURCE: Puzzanchera, C., Stahl, A. L., Finnegan, T. A., Snyder, H. N., Poole, R. S., & Tierney, N. (2000). *Juvenile Court Statistics 1997.* Washington, DC: National Center for Juvenile Justice.

2001, the number of girls arrested increased by 18.8%, whereas arrests of boys decreased by 9.2% (see Table 2.1 in Chapter 2). Population growth cannot explain these differences because boys' and girls' growth rates do not differ; the juvenile male and female populations have each grown by 10% since 1990.

In addition, the gendered nature of juvenile arrests also continues. As we saw earlier, although about four boys are arrested for every girl, the ratio for serious crimes of violence is about nine to one. Nearly 60% of those arrested in 2001 for running away were girls (FBI, 2002, p. 239). Looking at these figures differently, two status offenses (runaway and curfew violations) composed about 1 out of 5 arrests of girls but less than 1 in 10 arrests of boys. In addition, as noted in Chapter 2, the number of arrests of girls (and boys) for curfew violations has increased since 1992. Although arrests for runaways decreased between 1992 and 2001, girls' arrests for curfew violation increased by 56.7%, compared to a 25.6% increase for boys (FBI, p. 239). These arrest figures mean a considerable "front end" pressure on a juvenile justice system that has been told to "divert" and "deinstitutionalize" these youth.

What is also important to recognize is the racial and ethnic character of this "diversion" movement for the juvenile justice system. In 1997, for every 100,000 African American girls in the population, 234 were in juvenile custody, compared to 100 per 100,000 Hispanic girls and 75 per 100,000 white girls (Office of Juvenile Justice and Delinquency Prevention, 2001). In states where the custody rate of female juveniles exceeded the national average, the custody rate for girls of color was markedly higher (see Table 4.3). Wyoming, the state with the highest juvenile female custody rate, also led the nation with the largest proportion of girls in custody. In 1997, 43% of Wyoming's entire juvenile custody population were girls (Office of Juvenile Justice and Delinquency Prevention).

A growing number of studies have examined the development of a two-track juvenile justice system—one track for girls of color and another for white girls. In a study of investigation reports from one area office in Los Angeles, Jody Miller (1994) analyzed the effect of race and ethnicity on the processing of girls' cases during 1992–1993.

Comparing the characteristics of the youth in Miller's group with Schlossman's earlier profile of girls in Los Angeles in the 1950s shows how radically (and racially) different the current girls in the Los Angeles juvenile justice system are from their earlier counterparts. Latinas made up the largest proportion of the population (43%), followed by white girls (34%) and African American girls (23%; Miller, 1994, p. 11).

Predictably, girls of color were more likely to be from low-income homes, but this was especially true of African American girls (53.2% were from AFDC families, compared to 23% of white girls and 21% of Hispanic girls). Most important, Miller (1994) found that white girls were significantly more likely to be recommended for a treatment rather than a "detention-oriented"

Table 4.3   Female Juvenile Custody Rate (per 100,000 Female Juveniles), 1997

State With Highest Female Juvenile Custody Rate	Total	White	Black	Hispanic	American Indian/ Alaskan Native	Asian/ Pacific Islander
U.S. total	102	75	234	100	224	39
Wyoming	429	369	0	639	2,027	0
South Dakota	200	152	1,154	0	505	0
Nebraska	180	132	720	282	497	211
Kansas	177	137	575	207	203	224
Nevada	176	145	231	208	813	159
Alaska	171	101	600	197	344	167
Georgia	163	99	297	105	0	0
Louisiana	155	84	272	105	0	0
Indiana	150	126	356	186	0	0
Virginia	141	94	283	129	0	70
North Dakota	138	119	0	0	432	0

SOURCE: Office of Juvenile Justice and Delinquency Prevention. (2001). *OJJDP Statistical Briefing Book.* Retrieved December 15, 2002, from http://www.ojjdp.ncjrs.org/ojstabb/qa178.html

placement than either African American or Latina girls. In fact, 75% of the white girls were recommended for a treatment-oriented facility, compared to 34.6% of the Latinas and only 20% of the African American girls (p. 18).

Examining a portion of the probation officers' reports in detail, Miller (1994) found key differences in the ways that girls' behaviors were described—reflecting what she called "racialized gender expectations." In particular, African American girls' behavior was often framed as the product of "inappropriate 'lifestyle' choices," whereas white girls' behavior was described as the result of low self-esteem, being easily influenced, and "abandonment" (p. 20). Latina girls, Miller found, received "dichotomized" treatment, with some receiving the more paternalistic care white girls received and others receiving more punitive treatment (particularly if they committed "masculine" offenses, such as car theft).

Robinson (1990), in her in-depth study of girls in the social welfare (Child in Need of Supervision, or CHINS) and juvenile justice system (Department of Youth Services) in Massachusetts, documents the racialized pattern of juvenile justice quite clearly. Her social welfare sample ($N = 15$) was 74% white/non-Hispanic, and her juvenile justice system sample ($N = 15$) was 53% black or Hispanic.

Her interviews document the remarkable similarities of the girls' backgrounds and problems. As an example, 80% of the girls committed to DYS reported being sexually abused, compared to 73% of the girls "receiving services as a child in need of supervision" (Robinson, 1990, p. 311). The difference between these girls was in the offenses for which they were charged; all the girls receiving services were charged with traditional status offenses (chiefly running away and truancy), whereas the girls committed to DYS were charged with criminal offenses. Here, however, her interviews reveal clear evidence of bootstrapping. Take, for example, the 16-year-old girl who was committed to DYS for "unauthorized use of a motor vehicle." In this instance, "Beverly," who is black, had "stolen" her mother's car for 3 hours to go shopping with a friend. Prior to this conviction, according to Robinson's interview, she had been at CHINS for "running away from home repeatedly." "Beverly" told Robinson that her mother had been "advised by the DYS social worker to press charges for unauthorized use of a motor vehicle so that 'Beverly' could be sent to secure detention whenever she was caught on the run" (p. 202).

Other evidence of this pattern is reported by Bartollas (1993) in his study of youth confined in juvenile "institutional" placements in a midwestern state. His research sampled female adolescents in both public and private facilities. The "state" sample (representing the girls in public facilities) was 61% black, whereas the private sample was 100% white. Little difference, however, was found in the offense patterns of the two groups of girls. Seventy percent of the girls in the "state" sample were "placed in a training school as a result of a status offense" (p. 473). This state, like most, does not permit youth to be institutionalized for these offenses. However, Bartollas noted that "they can be placed on probation, which makes it possible for the juvenile judge to adjudicate them to a training school" (p. 473). In the private sample, only 50% were confined for status offenses; the remainder were there for "minor stealing and shoplifting-related offenses" (p. 473). Bartollas also noted that both of these samples of girls had far less extensive juvenile histories than did their male counterparts.

Other evidence, though less direct, points to much the same pattern. As deinstitutionalization has advanced over the past two decades, there has been a distinct rise in the number of youth confined in private and public "facilities" or institutions. In comparing "public" and "private" facilities, some clear gender and race differences emerge. Although the majority of the youths in all institutional populations are male, there is a noticeable gender difference between youth held in public and private facilities. In 1999, girls constituted 12% of those juveniles in public institutions, compared to 16% in private institutions (Office of Juvenile Justice and Delinquency Prevention, 2001; see also Moone, 1993a, 1993b, for earlier analysis). Girls were 7% more likely to go to private facilities than were boys.

There are also gender differences in the offenses or activities that bring youth to private facilities. In 1999, the majority of girls (63%) held in private facilities were being held for "nondelinquent" offenses, including status offenses, dependency and neglect, and "voluntary" admissions; for boys, only slightly less than a quarter (24%) were held for these reasons, with the rest being held for criminal offenses (Office of Juvenile Justice and Delinquency Prevention, 2001; see also Jamieson & Flanagan, 1989). In addition, girls were 1.5 times more likely than boys to be held for technical violations in private facilities (Office of Juvenile Justice and Delinquency Prevention).

An examination of California's juvenile justice system reveals this push for increased confinement that particularly affects girls. In 1999, California had the greatest number of juvenile females in custody—1926 girls, which is 13% of the entire nation's juvenile female custody population (Office of Juvenile Justice and Delinquency Prevention, 2001). Whereas nationally, the number of girls committed to correctional facilities increased 2% from 1997–1998 (and over half the states actually held fewer girls in 1999 than in 1997), California's juvenile female custody population increased by 8%, with a custody rate for African American girls that exceeded the national average— 320 per 100,000. Comparatively, boys' overall custody population decreased by 5% in California. Although California ranks fourth in the nation in juvenile offense rates at 549 per 100,000 juveniles, this cannot explain an increase in girls' confinement. Looking at Table 4.4, we can see that this increase for girls is not due to a growth in violence, drug crimes, or other felonies committed. Girls' murder and robbery arrests each decreased more than boys. Additionally, both boys' and girls' arrests for other felonies and drug violations also decreased (California Department of Justice, 1998, 1999). The only notable increase for girls and for boys was in the category of assaults, and even then the increase was not substantial (7% and 3%, respectively).

This increase in girls' custody populations in California may be due in large part to both the resistance of deinstitutionalization and diversion as well as the growing popularity of "tough on crime" policies. It is no coincidence then that by the year 2000, California began discussion of building a new

Table 4.4   California Juvenile Arrests by Sex, 1997–1999

Offense	Females			Males		
	1997–1998	*1998–1999*	*% Change*	*1997–1998*	*1998–1999*	*% Change*
Murder	22	12	−45	286	170	−41
Robbery	718	584	−19	6,103	5,128	−16
Assault	2,167	2,321	+7	9,938	10,261	+3
Drug	1,123	1,047	−7	6,269	5,541	−12
Other felony	11,580	10,849	−6	64,524	57,654	−10

SOURCE: California Department of Justice. (1998, 1999). *California Criminal Justice Profile.* Sacramento, CA: Division of Criminal Justice Information Services.

youth facility, with 90 new beds for boys and girls (Kirwan, 2000). Also troublesome are measures, such as Proposition 21, that embody the hyper-punitive approach toward juveniles in the criminal justice system: more juveniles to be tried as adults, mandatory confinement of certain juveniles in state or local correctional facilities, increased penalties for violent and serious offenses, and reduced confidentiality privileges. The impact of the potential fiscal costs (estimating to over one billion dollars) and the social effects of Proposition 21 have yet to be seen or extensively researched (Center on Juvenile and Criminal Justice, 2002).

With this push for confinement in states such as California comes ethnic and racial differences in the populations of these correctional institutions. Overall minority youth constitute 33% of the population age 10–17 in the United States, but they represent 63% of youth in residential placements (Office of Juvenile Justice and Delinquency Prevention, 2001). Whites constituted about 34% of those held in public institutions in 1997 but 46% of those held in private facilities (Office of Juvenile Justice and Delinquency Prevention). Among public juvenile correctional facilities, in 28% of facilities, minority youth comprised 75% and 100% of the offender population. Among private facilities, 21% had minority proportions in that range (Office of Juvenile Justice and Delinquency Prevention).

Other, more detailed work on this issue (Krisberg, Schwartz, Fishman, Eisikovits, & Guttman, 1986) suggests that there is a significant interaction between gender and ethnicity in incarceration rates. In addition, where increases have occurred in incarceration of youth, they have apparently been in the incarceration of minority youth, both male and female (Krisberg et al.,

1986). In 1982, for example, the ratio of black male incarceration rates to white male incarceration rates was 4.4 to 1. Hispanic males were incarcerated 2.6 times as often as white males. For girls, a similar though less extreme pattern was found, with black girls incarcerated 2.6 times as often as white females. A less dramatic difference was seen for Hispanic girls (only 1.1 times as often as white girls; pp. 16–17). More recently, in 1999, African American youth accounted for 39% of the juvenile custody population (46% of the male population and 34% of the female), while comprising only 14% of the population age 10–17 in the United States (Office of Juvenile Justice and Delinquency Prevention, 2001). In addition, minority youth spend more time in facilities than their white counterparts; minority youth spend an average of 17 weeks in custody while white youth spend an average of 15 weeks.

Other data comparing trends in detention between 1984 and 1988 found a 10.1% increase in detention of nonwhite girls for delinquency offenses, particularly drug offenses (where a fourfold increase in detentions was observed), compared to a drop in white girls' detentions for the same offenses (including drug offenses; Krisberg, DeComo, Herrera, Steketee, & Roberts, 1991, pp. 104–105). Turning to detention of girls for status offenses, a far more dramatic drop was seen in the detention of white girls for these offenses (30.5%) than for nonwhite girls (7.7%) during the same period.

In general, the numbers indicate that after a dramatic decline in the early 1970s, the number of girls held in public training schools and detention centers has not declined at all since 1979. Recent global data show this quite clearly. On one day in 1979, there were 6,067 girls in public facilities (mainly detention centers and training schools); in 1999, the figure had increased 56% to 9,444. Meanwhile, the number of girls held in private facilities had decreased by 51%—from 8,176 in 1979 to 5,073 in 1999 (Krisberg et al., 1991, p. 43; Moone, 1993a, 1993b; Office of Juvenile Justice and Delinquency Prevention, 2001).

OFFENSE PATTERNS OF
GIRLS IN CUSTODY—BOOTSTRAPPING

A slightly richer picture of the girls in correctional institutions comes from a national snapshot of the offenses for which girls and boys are held in public and private facilities. Tables 4.5 and 4.6 reveal that girls are still far more likely than boys to be held for status offenses in either type of facility and that

Table 4.5   Juveniles Detained in Public Facilities by Sex, 1999

Most Serious Offense	*Female (%)*	*Male (%)*
Delinquency	94	99
Person	32	37
Property	25	30
Drug	6	9
Public order	8	10
Probation/parole violation	23	12
Status offenses	6	1
Total in custody	**9,444**	**67,713**

SOURCE: Office of Juvenile Justice and Delinquency Prevention. (2001). *OJJDP Statistical Briefing Book*. Retrieved December 15, 2002, from http://www.ojjdp.ncjrs.org/ojstabb/qa178.html

Table 4.6   Juveniles Detained in Private Facilities by Sex, 1999

Most Serious Offense	*Female (%)*	*Male (%)*
Delinquency	76	93
Person	27	32
Property	22	30
Drug	7	10
Public order	6	11
Probation/parole violation	15	10
Status offenses	24	7
Total in custody	**5,073**	**26,526**

SOURCE: Office of Juvenile Justice and Delinquency Prevention. (2001). OJJDP Statistical Briefing Book. Retrieved December 15, 2002, from http://www.ojjdp.ncjrs.org/ojstabb/qa178.html

girls are far less likely to be held for violent offenses. As an example, 93% of the boys in private facilities are being held for delinquent offenses, compared to only 76% of the girls. Girls, in contrast, were often held for status offenses and "voluntary commitments," which take on a more sinister tone when one realizes that parents can "voluntarily" commit their children (Chesney-Lind & Shelden, 1992, p. 165).

The pattern persists in public detention centers and training schools, and another theme emerges. Essentially, 6% of the girls but only 1% of the boys were incarcerated for status offenses or were nonoffenders (see Table 4.5). Whereas boys were incarcerated for violent offenses or "serious" property offenses more often than girls (67% and 57%, respectively), girls were far more likely to be held for violations of probation or parole. Girls were nearly twice as likely to be held for technical violations than were boys. The large number of girls incarcerated for probation or parole violations is a measure of new efforts to bootstrap status offenders into delinquents by incarcerating them for these offenses (Girls Incorporated, 1996, p. 7).

Such new efforts to bootstrap status offenses into delinquent charges (for example, assault or technical violations) becomes apparent when changes over time in the female juvenile custody population are analyzed. In 1999, 9,444 girls were committed to public correctional facilities and 5,073 were committed to private ones. Compared to the female population in custody in 1991, this represents a 49% increase in the public facilities and a 51% decrease in private ones (Girls Incorporated, 1996; Office of Juvenile Justice and Delinquency Prevention, 2001). The percentage of nonoffenders and voluntary commitments in private facilities dropped from 65% of girls in private facilities in 1991 to 24% in 1999. Whereas status offenders were 19% of females in custody in 1991, they were only 13% in 1999. Furthermore, person offenders comprised only 8% of females in custody in 1991; in 1999, they comprised 30% of girls in custody.

One explanation for this decline in voluntary commitments and nonoffenders in girls' private facilities lies in changes within the health care system. With the advent of health management organizations (HMOs), the approval for funding adolescents with "behavioral" problems became more difficult to achieve (see Pasko, 1997, for this discussion). Refusing payment for inpatient or residential treatment, HMOs shifted to intensive outpatient and family therapy for girls with "conduct problems." Parents could no longer voluntarily commit their "incorrigible" daughters with ease of approval from their health care insurances. Given the cost of inpatient treatment and the limited assistance from the health care insurance, parents turned to the juvenile justice system to institutionalize their daughters. In order to be committed via the justice system, girls' behaviors (such as fighting with a parent, running away, staying out after curfew) needed to be "upcrimed' and relabeled as delinquent offenses. This bootstrapping phenomena explains both girls' increases in person offenses and in the custody populations in public facilities as well as their decrease in status offenses and in private institutions' populations.

A study of girls committed to or in state care in Virginia further underscores this concern to focus on bootstrapping. The study found that minority girls, who comprise about 26% of Virginia's population, make up fully half of

Table 4.7   Virginia Secure Custody: Probation Office Reasons for
                   Recommending Commitment

Reasons for Commitment	Frequency
Probation violation	46
Repeated runaway	44
Self-victimization	43
Failure to participate in ordered treatment/service	29
Chronic delinquency	27
Punishment	8
Treatment not available locally	7
Heinous violent crime	7
Frustration/exasperation within system	5
Noncompliance with court-ordered fine, restitution, public service	5
Example to community/other juveniles	5

SOURCE: Task Force on Juvenile Females Offenders. (1991). *Young women in Virginia's juvenile justice system: Where do they belong?* Richmond, VA: Department of Youth and Family Services.

those committed to a secure facility. The most frequent offenses committed by these girls were misdemeanor offenses, followed by status offenses. Most revealing, though, were the responses of the probation counselors to the request that they rank the reason for the girls' commitments (see Table 4.7). The most common answer was "probation violation," followed by "repeated runaway," "self-victimization," and "failure to participate in ordered treatment/ service." Far down the list was "heinous violent crime" and "punishment" (Task Force on Juvenile Female Offenders, 1991, pp. 2–3).

These patterns have also turned up in other countries. Reitsma-Street (1993) found that in Canada in 1991, "one in four charges laid against young females are against the administration of justice; the rate is one in six for males" (p. 445). In Canada, violation of a court order or failure to comply with decisions of the youth administration of justice are described as "offenses against the administration of justice," or, stated more simply, bootstrapping, Canadian style.

Taken together, the data on girls currently being held in public institutions show tremendous and high-level resistance to the notion that youth who have not committed any criminal act should not be held in institutions. Although the juvenile female population in private facilities has decreased over the years, girls are still 3.5 times more likely than boys to be nonoffenders or voluntary commitments in these institutions. Having said this, it would be remiss not to note that the deinstitutionalization movement has reduced the number of girls in detention centers and training schools. The movement may, however, have simply moved those girls into a private system of institutionalization—the mental health system.

DEINSTITUTIONALIZATION OR TRANSINSTITUTIONALIZATION? GIRLS AND THE MENTAL HEALTH SYSTEM

Despite considerable resistance, the incarceration of young women in public training schools and detention centers across the country fell dramatically after the passage of the Juvenile Justice and Delinquency Prevention Act of 1974. Prior to the passage of the act, nearly three quarters (71%) of the girls and 23% of the boys in the nation's training schools were incarcerated for status offenses (Schwartz, Steketee, & Schneider, 1990). Between 1974 and 1979, the number of girls admitted to public detention facilities and training schools dropped by 40%. Since then, however, the deinstitutionalization trend has slowed in some areas of the country, particularly at the detention level, as the number of girls held in public facilities has increased since 1979 (Moone, 1993b; Office of Juvenile Justice and Delinquency Prevention, 2001; U.S. Department of Justice, 1989, p. 43).

In addition, there has been considerable gender disparity in the commitment of girls to private facilities—8% of the entire girls' custody populations are voluntary commitments or nonoffenders, compared to 2% for boys (Moone, 1993a; Office of Juvenile Justice and Delinquency Prevention, 2001; U.S. Department of Justice, 1989, p. 43). Although some of this is doubtless good news (at least for white girls), others have taken a more critical look at this trend. Schwartz, Jackson-Beeck, and Anderson (1984) have called this a system of "hidden," private juvenile correction in which incarceration can occur without any legal procedure and without the consent of the youths (because they are underage). Costs are covered by third-party health care insurance plans; indeed the reliance on this funding structure is one reason that the trend has slowed. Health care plans, under new pressure to "manage" costs, began to take a hard look at this practice and began to prohibit reimbursement for conduct disorders (see Chesney-Lind and Pasko, 2003).

The clearest problems with private institutions arise in the case of private psychiatric hospitals. Although the number of juveniles institutionalized for inpatient psychiatric conditions went down in the latter half of the 1990s, this has not always been the case. Between 1980 and 1984, adolescent admissions to psychiatric units of private hospitals increased fourfold (Weithorn, 1988, p. 773). There had also been a marked shift in the pattern of juvenile mental health incarceration. In 1971, juvenile admissions to private hospitals accounted for 37% of all juvenile admissions, but by 1980 this figure had risen to 61% (p. 783). Finally, it is estimated that fewer than one-third of juveniles admitted for inpatient mental health treatment were diagnosed as having severe or acute mental disorders, in contrast to between half and two thirds of adults admitted to these sorts of facilities (pp. 788–789). A closer look by Weithorn at the youth in such institutions in Virginia suggests that 36% to 70% of the state's hospital population "suffer from no more than 'acting out' problems and a range of less serious difficulties" (p. 789). Despite this, juvenile psychiatric patients remain in the hospital approximately twice as long as adults (p. 789).

These patterns have very clear implications for the treatment of girls, though data on the gender of patients is regrettably sketchy. The chief concern is that admissions procedures to such private facilities have never been formalized. In addition, the Supreme Court of the United States has clearly rejected an attempt to guarantee procedural protections to juveniles that have been extended to adults in Parham v. J. R. (1979). Perhaps as a consequence of both the Supreme Court's decision and the partial success of deinstitutionalization within the juvenile justice system, many private hospitals, whose profits are enhanced by filling beds, advised institutional care for "troublesome" youth (Weithorn, 1988, p. 786). Some of the criteria generated by these institutions were of specific concern to those worried about the overuse of institutionalization. For example, one set of criteria suggested that "sexual promiscuity" is an example of a "self-defeating" or "self-destructive behavior" necessitating "immediate acute-care hospitalization [as] the only reasonable intervention" (p. 786). According to Weithorn, "inability to function" was another cause of incarceration. This phrase included an inability to function in the following areas: family life, vocational pursuits, and "choice of community resources." This last phrase was defined as "relating to vocational interests in school, church activities, scouting activities, the expression of hobbies and/or special interest in the community, as well as the individual's choice of peers for nonstructured community activities" (p. 786).

Given this vague language and the problems that public systems of control have had with sexist interpretations of status offense labels, it should come as no surprise that a number of the examples of egregious abuse of institutionalization provided by Weithorn (1988) involved girls. Some of these cases made the link directly between status offenses and these incarcerations. The case of "Sheila," for example, involved a 12-year-old girl who was hospitalized in a state psychiatric facility after spending a week in a juvenile detention center on the basis that she was a "child in need of supervision." In another case ("Lisa"), a 16-year-old girl was admitted into a private psychiatric hospital because she "'seduced' older men, drank vodka, skipped school, ran way from home, and disobeyed her divorced mother" (p. 790). Ironically, as noted earlier, it was not a concern with the rights of girls that has slowed this practice of institutionalizing girls in private facilities for these sorts of behaviors, but instead a desire to contain costs in medical care during a period of managed care.

Other privately funded programs should also be scrutinized carefully for evidence of gender bias and abuse. An extreme example of abuse turned up in a "tough love" type of boot camp in Colorado. Youth complained that counselors "spit in their faces, made them eat their own vomit, challenged them to fight, screamed racial and sexist slurs at them and made them carry human feces in their pockets" (Weller, 1996, p. 1). Investigators discovered the abuse when two youth were found to have a flesh-eating virus and a girl lost a finger. Youth spent "between two and 12 weeks" at the camp because "their parents had problems with them and wanted them in a regimented environment" (p. 2).

In another case, Mystie Kreimer, a 15-year-old resident of a 161-bed youth "shelter and treatment center" died en route to a hospital. Mystie had been placed as a "resident" in one of a "fast-growing chain of for-profit congregate care children's facilities" (Szerlag, 1996, p. 48). Started by a politically connected multimillionaire who launched Jiffy Lube, Youth Services International now operates such homes in 10 states. The facility that held Mystie, Forest Ridge, had been investigated by the Iowa Department of Human Services after former staff reported that "teens were frequently physically and emotionally abused by inexperienced youth workers." In addition, a former staff member plead guilty to "sexually exploiting a minor" (p. 48).

Mystie landed at Forest Ridge because she had a "history of substance abuse and sexual adventuring" (Szerlag, 1996, p. 42). She entered the facility in March 1995, weighing 145 pounds. When she was visited by her mother in September of that same year, she complained of leg and chest pains. Her mother recalls asking to have her daughter hospitalized but was told by the facility that she was not "sick enough to warrant that kind of medical attention" (p. 42). The mother of another girl reports seeing Mystie lying on a couch and commenting to her daughter that Mystie looked sick. Her daughter replied, "Oh, Mom, she is so sick and they still expect her to do her chores" (p. 42). Mystie died while being transported by helicopter to a Sioux Falls hospital. At the autopsy, it was determined that she died of a "massive blood clot in her lung." At the time of her death, she weighed 100 pounds (p. 42).

Although the data are far from complete, the evidence seems to indicate that girls, particularly middle-class white girls, were being incarcerated in private hospitals and treatment programs for much the same behavior that, in previous decades, placed them in public institutions. Given the lack of procedural safeguards in these settings, some might even argue that there is greater potential for sexist practices and even abuse to flourish in these closed and private settings. Although girls are still "voluntarily" committed to private institutions more often than boys, the frequency of this appears to be changing since the late 1990s. Perhaps an unintended consequence from managed care is the beginning cessation of voluntary commitments for girls with "conduct" problems. This unintended consequence, however, may very well mean a reversal to the practices of previous decades—more girls in public institutions.

SMALL NUMBERS DON'T MEAN
SMALL PROBLEMS: GIRLS IN INSTITUTIONS

In some ways, as the numbers of girls decreased, the small amount of public attention that had been devoted to girls' problems in institutions has disappeared. This does not mean, however, that the problems associated with the institutionalization of girls have magically disappeared. There are still girls in institutions today, and their small numbers have begun to produce for them the sorts of problems that once characterized the women's prison populations that were correctional afterthoughts in systems devised to control men. One student of this pattern of institutional neglect noted that girls in institutions have become the "forgotten few" (Bergsmann, 1989). In addition, girls' small numbers can become even more of a problem in small town and rural areas, where the risks for serious neglect can be more profound.

Institutional practices that are routine in boys' facilities, for example, must be scrutinized for their effect on girls. In particular, there needs to be sensitivity to the fact that girls' victimization histories make such practices as routine strip searches and isolation extremely risky. There should be no surprise that scandals routinely occur when girls are strip searched. Take the case of a 16-year-old runaway who was picked up on a curfew violation in San Francisco. She initially refused to allow a strip search because she "didn't want to take her clothes off in front of a man." She was then "forcibly strip searched," during which a male staff member "helped hold the girl down on the floor while a female staffer removed the girl's clothes" (Goldberg, 1996, p. A1).

Problems also exist in public facilities that have undergone downsizing, where the small number of girls housed is frequently used to deny them access to services and programs. Like adult women inmates in other decades, the small number of girl inmates invites new forms of abuse and neglect. As illustrated in the headlines that opened this chapter, the potential for sexual assault, abuse, and mistreatment in girls' detention centers remains problematic.

A clear example of these problems surfaced in the year-long investigation of the Lloyd McCorkele Training School for Boys and Girls in Skillman, New Jersey. The training school officially closed in 1992 and all the male inmates and most of the staff were moved out of the facility. Correctional planners were not able to locate a new facility for the girls' unit so, 17 months later, the facility continued to house about a dozen girls (Rimbach, 1994, p. 32).

Three girls who agreed to talk to a reporter told stories of sexual assault by the male correctional staff (two formal complaints were lodged), beatings, extensive use of isolation, and overreliance on "mind-numbing" medications. One girl spoke of spending 8 months in a "stark isolation cell" for infractions such as "cursing at correctional officers," "refusing to go to bed," or speaking to another girl in the next cell. The facility was also routinely transferring girls who turned 18 to the adult woman's prison, whereas the boys' training school kept boys until age 21 (Rimbach, 1994, p. 34).

Journalists investigating the facility also reported extensive use of psychotropic medication; roughly one third of all the girls who went through the facility in the last year had been given medication, compared to no more than about 4.5% of the boys (Rimbach, 1994, p. 31). A letter from one of the advisory board members who visited the facility in 1992 provides graphic details of the conditions in the girls' facility:

> During my visit, the girls looked extremely unkempt, they were excessively lethargic, their attitude was very poor, reportedly the majority were on some kind of medication, either tranquilizers or sleeping pills, and four were in "administrative segregation" without even a book to read.
>
> Additionally, there were allegations that these girls were not allowed out for the required one hour a day. I was told that one young woman had not had a shower for five days.
>
> While I am no expert on juvenile law, I believe that there are several violations going on in this facility. These girls are not being given equal treatment to the boys in the system and they are not being given the physical education required by law. Further, as an enlightened administrator, you should be shocked at the total absence of program in this facility, particularly as there is specific funding for this program. (Rimbach, 1994, p. 31)

When asked about these problems, correctional administrators provided an interesting litany of excuses. About the overmedication, administrators noted that "girls come into Skillman from county detention centers already on medication." They said that girls "usually get so many chances before they are committed, so those who reach Skillman are a pretty disturbed bunch of kids" (Rimbach, 1994, p. 31). Regarding the extensive use of isolation (correctional records showed that each of the eight girls held in 1993 spent an average of 30 days in isolation), the administrator could only suggest that perhaps the girls had come from other facilities to be kept in isolation at Skillman. Yet the record shows that the small facility averaged 10 disciplinary charges a month. The absence of programming was explained by saying that "the girls have been in transition" (p. 31).

Unfortunately, detailed accounts of youth facilities such as this one have been extremely scarce in the past few decades. Earlier work (see Chesney-Lind & Shelden, 1992) found different and less serious abuses against girls held in large facilities during the 1960s and 1970s. Clearly, as the number of girls in public facilities declines, the girls are forgotten in a system designed to hold young men. With that "invisibility," as the Skillman school and the San Francisco Youth Guidance Center demonstrate, also comes the possibility for extreme correctional neglect and abuse.

INSTEAD OF INCARCERATION: WHAT COULD BE DONE TO MEET THE NEEDS OF GIRLS?

Girls on the economic and political margins, particularly those who find their way into the juvenile justice system, share many problems with their male counterparts. They are likely to be poor, from disrupted and violent families, and having trouble in school. In addition, however, girls also confront problems unique to their sex: notably sexual abuse, sexual assault, dating violence, depression, unplanned pregnancy, and adolescent motherhood. Their experience of the problems they share with boys and the additional problems they face as girls are both conditioned by their gender, class, and race. Because families are the source of many of the serious problems that girls face, solutions must take into account the possibility that some girls may not be able to stay safely at home.

Programming for girls clearly needs to be shaped by girls' unique situations and to address the special problems girls have in a gendered society. Unfortunately, traditional delinquency treatment strategies, employed in both prevention and intervention programs, have been shaped largely by common-sense assumptions about what youth—generally boys—need, and even then, these problems fail to recognize boys' gender management strategies and problematic dimensions. Sometimes girls will benefit from these notions, and sometimes their problems will not be addressed at all.

There is a tremendous shortage of information on programs that have been proven effective with girls (see Chesney-Lind & Shelden, 1992). Indeed, many studies that have evaluated particular approaches do not deal with special gender issues and frequently programs do not even serve girls. In addition, programs that have been carefully evaluated are often set in training schools (clearly not the ideal place to try any particular strategy). Finally, careful evaluation of most programs shows that even the most determined efforts to intervene and help often have very poor results. Of course, the last two points may be related; programs set in closed, institutional settings are clearly at a disadvantage and, as a consequence, tend to be less effective (Lipsey, 1992). Unfortunately, community-based programs for girls have been few and far between.

Because of this, it might be useful to briefly review the general problems associated with the establishment of programs for girls. Prior to the 1970s (and the second wave of feminism), there was little concern over uniquely female problems, such as wife battering and sexual assault, and even less concern about girls' problems (Gordon, 1988). As we have seen, instead of girls' victimization, it was girls' sexuality that was perceived as the problem. The typical response to girls' sexuality was first to insist that the girls return to their dysfunctional families. If they ran from these settings, they were placed in a foster home (if they were lucky) or, ultimately, an institution, "for their own protection."

After two decades of deinstitutionalization efforts, girls remain "all but invisible in programs for youth and in the literature available to those who work with youth" (Davidson, 1983, p. viii). Furthermore, programs for young women in general (and delinquents in particular) have been of low priority in our society, as far as funding is concerned. For instance, a report written in 1975 by the Law Enforcement Assistance Administration revealed that only 5% of federally funded juvenile delinquency projects were specifically directed at girls, and that only 6% of all local monies for juvenile justice were spent on girls (Female Offender Resource Center, 1977, p. 34). A more recent review of 75 private foundations revealed that funding "targeted specifically for girls and women hovered around 3.4 percent" (Valentine Foundation and Women's Way, 1990, p. 5).

An exhaustive study of virtually all program evaluation studies done since 1950 (Lipsey, 1992) located reports about some 443 delinquency programs; of these, 34.8% exclusively served boys and an additional 42.4% served "mostly males." Lipsey found that only 2.3% of the surveyed programs explicitly served only girls, and only 5.9% of the programs served "some males," meaning that most of the programs' participants were girls (p. 106).

Finally, a 1993 study of the San Francisco Chapter of the National Organization for Women found that only 8.7% of the programs funded by the major city organization funding children and youth programs "specifically addressed the needs of girls" (Siegal, 1995, p. 18). Not surprisingly, then, a 1995 study of youth participation in San Francisco after-school or summer sports programs found that only 26% of the participants were girls (Siegal, p. 20).

What are the specific needs of young women; in particular, those who come into contact with the juvenile justice system as victims or offenders? Davidson (1983) argues the following:

> The most desperate need of many young women is to find the economic means of survival. While females today are still being socialized to believe

that their security lies in marriage and motherhood, surveys of teenage
mothers indicate that approximately 90 percent receive no financial aid from
the fathers of their children. (p. ix)

Likewise, a study of homeless youth in Waikiki, about half of whom were
girls, revealed that their most urgent needs are housing, jobs, and medical
services (Iwamoto, Kameoka, & Brasseur, 1990). Finally, a survey conducted
in a very poor community in Hawaii (Waianae) revealed that pregnant and
parenting teens saw medical care for their children, financial assistance, and
child care as their major needs. Social workers in the same community, in contrast, saw parenting classes as the girls' most important need, followed by
child care, educational and vocational training, and family planning (Yumori
& Loos, 1985, pp. 16–17). These findings suggest that although youth understand that economic survival is their most critical need, such is not always the
case for those working with them.

The Minnesota Women's Fund noted that the most frequent risk factors
for girls and boys differ, and that for girls the list includes emotional stress,
physical and sexual abuse, negative body image, disordered eating, suicide,
and pregnancy. For boys, the list included alcohol, polydrug use, accidental
injury, and delinquency (Adolescent Female Subcommittee, 1994). Although
not all girls at risk will end up in the juvenile justice system, this gendered
examination of youth problems sets a standard for the examination of
delinquency prevention and intervention programs.

Among other needs that girls' programs should address are dealing with the
physical and sexual violence in their lives (from parents, boyfriends, pimps, and
others); confronting the risk of AIDS; dealing with pregnancy and motherhood;
countering drug and alcohol dependency; facing family problems; obtaining
vocational and career counseling; managing stress; and developing a sense of
efficacy and empowerment. Many of these needs are universal and should be
part of programs for all youth (Schwartz & Orlando, 1991). However, most of
these are particularly important for young women.

Alder (1986, 1995) points out that serving girls effectively will require
different and innovative strategies because "young men tend to be more
noticeable and noticed than young women" (Alder, 1995, p. 3). When girls go
out, they tend to move in smaller groups, there are greater proscriptions
against girls "hanging out," and they may be justly fearful of being on the
streets at night. Finally, girls are subject to many more domestic expectations

than their male counterparts and these may keep them confined to their homes. Alder notes that this may be a particular issue for immigrant girls.

Despite the lack of evaluation research and the obvious necessity to recognize the special and unique needs of girls, better programming has emerged within the last decade. The Office of Juvenile Justice and Delinquency has been identifying (and funding) programs around the country that are working successfully with girls (Center for Juvenile and Criminal Justice, 2002, p. 2). Table 4.8 is a brief summary of just a few residential, day treatment, and community-based initiatives. Such programs work to address the aforementioned problems by valuing the female perspective, celebrating and honoring the female experience, taking into account female development, empowering girls to reach their potential, and working to change attitudes that might prevent girls from reaching such potential (Center for Juvenile and Criminal Justice; Office of Juvenile Justice and Delinquency Prevention, 1998).

There is also some encouraging news as girl-serving organizations (like the YWCA, Girls Incorporated, etc.) realize they have a responsibility for girls who are in the juvenile justice system. Recent reviews of promising programs for girls (Girls Incorporated, 1996; Schwartz & Orlando, 1991) indicate that programs that specifically target the housing and employment needs of youth, while also providing them with the specific skills they need to survive on their own, are emerging. These often include built-in caseworker/service broker and counseling components. Clearly, many girls will require specialized counseling to recover from the ravages of sexual and physical victimization, but the research cautions that approaches that rely simply on the provision of counseling services are not likely to succeed (see Chesney-Lind & Shelden, 1992).

Programs must also be scrutinized to assure that they are culturally specific. As increasing numbers of girls of color are drawn into the juvenile justice system (and bootstrapped into correctional settings) while their white counterparts are deinstitutionalized, there is a need for programs rooted in specific cultures. Because girls of color have different experiences of their gender and different experiences with the dominant institutions in the society (Amaro, 1995; Amaro & Agular, 1994; LaFromboise & Howard-Pitney, 1995; Orenstein, 1994), programs to divert and deinstitutionalize must be shaped by the unique developmental issues confronting minority girls, and must build in the specific cultural resources available in ethnic communities. In addition to programs listed in Table 4.8 that target minority girls, programs such as Diineegwasii in Fairbanks, Alaska (Alaskan Native girls), and Nuevo

Table 4.8   Examples of Promising Gender-Specific Programming Initiatives

Program	Location	Program Setting	Age Group/Ethnicity	Specific Gender-Related Programming	Other Specific Targeted Factors
Alternative Rehabilitation Communities (ARC)	Harrisburg, PA	Residential Continuum of Care	Ages 15-18 Predominantly African American	• Relationship building • Victimization • Nontraditional vocational training	• Parenting training • Female sex offenders
Caritas House	Pawtucket, RI	Residential	Ages 13-17 Caucasian	• Sexual abuse • Victimization • Relationship building • Staff training	• Alcohol and drug treatment facility
PACE Center for Girls	Headquarters Jacksonville, FL	Day Treatment	Ages 12-18 Caucasian and African American	• Relationship building • Staff training • Life skills • Positive gender identity	• Small all-girls' classes • Community service
Harriet Tubman Residential Center	Auburn, NY	Residential	Ages 15-18 Caucasian, African American, and Latina	• Relationship building • Staff training • Women's studies curriculum • Victimization • Self-empowerment skills • Positive gender identity	
G.I.R.L.S. on the Move!	Boston, MA	Community-based	Ages 10-16 African American and Latina/Hispanic	• Relationship building • Mentoring program • Life skills	• Alcohol and drug prevention • Entrepreneur program
Naja Project	Washington, DC	Community-based	Ages 10-14 African American	• Relationship building • Life skills • Leadership development	• Alcohol and drug prevention • Positive ethnic identity

SOURCE: Adapted from Office of Juvenile Justice and Delinquency Prevention. (1998). *Guiding Principles for Promising Female Programming (appendix)*. Washington, DC: Author.

Dia in Salt Lake City, Utah (Hispanic girls), are two more examples of gender-specific programming designed for minority girls. Each program works to develop both a positive gender as well as ethnic identity (Office of Juvenile Justice and Delinquency Prevention, 1998).

Innovative programs must also receive the same sort of stable funding generally accorded their more traditional and institutional counterparts (which are generally far less innovative and flexible). Many novel programs relied on federal funds or private foundation grants; pitifully few survived for any length of time. To survive and thrive, innovative programs must be able to count on stable funding. The recent pressure exerted by Congress on states to conduct an inventory of programs that are specifically focused on girls may pressure those receiving federal funding, in the words of Abigail Adams over two centuries ago, to "remember the ladies" (Rossi, 1973, pp. 10–11).

Finally, programs must be continually scrutinized to guarantee that they are serving as genuine alternatives to girls' incarceration, rather than simply extending the social control of girls. There is a tendency for programs serving girls to become "security" oriented in response to girls' propensity to run away. A component of successful programming for girls must be advocacy and the continuous monitoring of closed institutions. If nothing else can be learned from a careful reading of the rocky history of nearly two decades of efforts to decarcerate youth, one must appreciate how fraught with difficulty these efforts are and how easily their gains can be eroded.

Finally, much more work needs to be done to support the fundamental needs of girls on the margin. We must do a better job of recognizing that these girls need less "programming" and more support for living on their own, because many cannot or will not be able go back home again. Without financial, emotional, and psychological support or a place to call home, these girls could very well return to the criminal justice system—this time as women offenders.

NOTES

1. This phrase is taken from an inmate file by Rafter (1990) in her review of the establishment of New York's Albion Reformatory.

2. Recently, the juvenile justice reforms signaled in these hearings have been challenged by congressional initiatives that make it easier to detain girls for status offenses and to eliminate the small amount of money set aside for girls' programs (Howard, 1996).

FIVE

TRENDS IN WOMEN'S CRIME

———•◦•———

W omen's crime, like girls' crime, is deeply affected by women's place. As a result, women's contribution to serious and violent crime—like that of girls—is minor. Of those adults arrested for serious crimes of violence in 2001 (murder, forcible rape, robbery, and aggravated assault), only 17% were female. Indeed, women constituted only 21.4% of all arrests during that year (FBI, 2002, p. 239). This also means that adult women are an even smaller percentage of those arrested than their girl counterparts (who now comprise more than one out of four juvenile arrests).

Moreover, the majority of adult women offenders, like girls, are arrested and tried for relatively minor offenses. In 2001, women were most likely to be arrested for larceny theft (which alone accounted for 12.1% of all adult women's arrests), followed by drug abuse violations (10.1%). This means that more than a fifth of all the women arrested in the United States that year were arrested for one of these two offenses. Women's offenses, then, are concentrated in just a few criminal categories, just as women's employment in the mainstream economy is concentrated in a few job categories. Furthermore, these offenses, as we shall see, are closely tied to women's economic marginality and the ways women attempt to cope with poverty.

UNRULY WOMEN: A BRIEF
HISTORY OF WOMEN'S OFFENSES

Women's concentration in petty offenses is not restricted to the present. A study of women's crime in 14th-century England (Hanawalt, 1982) and descriptions of the backgrounds of the women who were forcibly transported to Australia several centuries later (Beddoe, 1979) document the astonishing stability in patterns of women's lawbreaking.

The women who were transported to Australia, for example, were servants, maids, or laundresses convicted of petty theft (stealing, shoplifting, and picking pockets) or prostitution. The number of women transported for these trivial offenses is sobering. Between 1787 and 1852, no less than 24,960 women, fully a third of whom were first offenders, were sent to relieve the "shortage" of women in the colonies. Shipped in rat-infested holds, the women were systematically raped and sexually abused by the ships' officers and sailors, and the death rate in the early years was as high as one in three. Their arrival in Australia was also a nightmare; no provision was made for the women and many were forced to turn to prostitution to survive (Beddoe, 1979, pp. 11–21).

Other studies add different but important dimensions to the picture. For example, Bynum's (1992) research on "unruly" women in antebellum North Carolina adds the vital dimension of race to the picture. She notes that the marginalized members of society, particularly "free black and unmarried poor white women" were most often likely both to break social and sexual taboos and to face punishment by the courts. Indeed, she observes that "if North Carolina lawmakers could have done so legally, they would have rid society altogether" of these women (p. 10). As it was, they harshly enforced laws against fornication, bastardy, and prostitution in an attempt to affect these women and their progeny.

The role of urbanization and class is further explored in Feeley and Little's (1991) research on criminal cases appearing in London courts between 1687 and 1912, and Boritch and Hagan's (1990) research on arrests in Toronto between 1859 and 1955. Both of these studies examine the effect of industrialization and women's economic roles (or economic marginalization) on women's offenses. Both works present evidence that women were drawn to urban areas, where they were employed in extremely low-paid work. As a result, this forced many into forms of offending, including disorderly conduct, drunkenness, and petty thievery. Boritch and Hagan make special note of the large numbers of women arrested for property offenses, "drunkenness," and "vagrancy," which can be seen as historical counterparts to modern drug offenses. But what of women who committed "serious" offenses such as murder? Jones's (1980) study of early women murderers in the United States reveals that many of America's early women murderers were indentured servants. Raped by calculating masters who understood that giving birth to a "bastard" would add 1 to 2 years to a woman's term of service, these desperate women hid their pregnancies and then committed infanticide. Jones also provides numerous historical and contemporary examples of desperate women murdering their brutal "lovers" or husbands. The less dramatic links between forced marriage, women's circumscribed options, and women's decisions to kill, often by poison, characterized the Victorian murderesses. These women, though rare, haunted the turn of the century, in part because women's participation in the methodical violence involved in arsenic poisoning was considered unthinkable (Hartman, 1977).

In short, research on the history of women's offenses, and particularly women's violence, is a valuable resource for its information on the level and character of women's crime and as a way to understand the relationship between women's crime and women's lives. Whenever a woman commits murder, particularly if she is accused of murdering a family member, people immediately ask, "How could she do that?" Given the enormous costs of being born female, that may well be the wrong question. The real question, as a review of the history of women's crime illustrates, is not why women murder but rather why so few murder.

Take a look at some facts. Every 15 seconds, a woman is beaten in her own home (Bureau of Justice Statistics, 1989). One in every three women report having been physically attacked by an intimate partner some time in her life (Wilt & Olson, 1996). In 1999, women were three times more likely to be killed by their intimate partners than were men and accounted for 85% of all victims of domestic violence (Rennison, 2001, p. 1). A National Institute of Mental Health study (based on urban area hospitals) estimated that 21% of all women using emergency surgical services had been injured in a domestic violence incident; that half of all injuries presented by women to emergency surgical services occurred in the context of partner abuse; and that over half of all rapes of women over the age of 30 had been perpetrated by an intimate partner (Stark et al., 1981). In addition, other studies have shown that marital rape is often more violent and repetitive than other forms of sexual assault and is often not reported (Richie, 2000, p. 4). In the United States, for example, former Surgeon General C. Everett Koop (1989) estimated that 3 to 4 million women are battered each year; roughly half of them are single, separated, or divorced (Rennison). According to the National Crime Victimization Survey, women separated from their husbands are victimized at higher rates than married, divorced, or single women, with females aged 16 to 24 experiencing the highest rate of intimate partner violence—151 per 1,000 women (Rennison, p. 5). Battering also tends to escalate and become more severe over time. Almost half of all batterers beat their partners at least three times a year (Straus, Gelles, & Steinmetz, 1980). This description of victimization doesn't address other forms of women's abuse, such as incest and sexual assault, which have rates as alarmingly high (see Center for Policy Studies, 1991).

The real question is why so few women resort to violence in the face of such horrendous victimization—even to save their lives. In 2001, in the United States, only 14.2% of those arrested for murder were women—meaning that murder, like other forms of violent crime, is almost exclusively a male activity. In fact, women murderers, as both Jones and Hartman document, are interesting precisely because of their rarity. The large number of women arrested for trivial property and moral offenses, coupled with the virtual absence of women from those arrested for serious property crimes and violent crimes, provides clear evidence that women's crime parallels their assigned role in the rest of society (Klein & Kress, 1976). In essence, women's place in the legitimate economy largely relegates them to jobs that pay poorly and are highly sex segregated (such as secretarial and sales jobs). Likewise, in the illicit or criminal world, they occupy fewer roles and roles that do not "pay" as well as men's crime. There is, however, little understanding of why this is the case and, until recently, little scholarship devoted to explaining this pattern. This chapter attempts to address both the reality of women's crime and the fascination with the atypical woman offender who is violent and defies her "conventional role" in both the mainstream and the criminal world.

TRENDS IN WOMEN'S ARRESTS

Over the years, women have typically been arrested for larceny theft, drunk driving, fraud (the bulk of which is welfare fraud and naive check forgery), drug abuse violations, and buffer charges for prostitution (such as disorderly conduct and a variety of petty offenses that fall under the broad category of "other offenses"; Steffensmeier, 1980; Steffensmeier & Allan, 1995; see Table 5.1).

Table 5.1   Adult 10-Year Trends by Sex, 1992–2001

Offense Charged	Men	1992 % Change	Women	1992 % Change
Total	4,829,151	−2.6	1,318,566	+15.0
Index Offenses:				
Murder and nonnegligent manslaughter	5,223	−32.0	865	−4.0
Forcible rape	11,726	−29.0	143	−22.0
Robbery	38,524	−24.8	4,592	−9.4
Aggravated assault	182,362	−8.7	43,668	+33.3
Burglary	62,001	−55.3	16,070	−0.01
Larceny-theft	284,265	−25.3	160,003	−18.2
Motor vehicle theft	41,673	−18.9	8,010	+38.5
Arson	4,159	−10.1	993	+22.6
Total violent crime	237,835	−13.6	49,268	+26.6
Total property crime	424,369	−26.2	185,076	−15.2
Other Offenses:				
Other assaults	445,832	+8.2	124,704	+57.7
Forgery and counterfeiting	36,603	+12.2	24,688	+37.7
Fraud	90,496	−16.3	78,085	−15.8
Stolen property: buying, receiving, possessing	43,052	−23.2	9,063	+3.0
Offenses against family	58,271	+16.7	15,547	+84.6
Prostitution, vice	13,484	−21.0	23,118	−18.5
Embezzlement	5,186	+18.3	5,359	+78.9
Vandalism	72,936	−11.0	15,888	+23.0
Weapons (carrying, etc.)	62,745	−31.5	5,136	−30.3
Drug abuse violations	585,206	+31.4	133,006	+42.3
Gambling	2,344	−55.9	405	−48.8
Liquor law violations	206,040	+18.6	55,951	+55.5
Driving under the influence	653,307	−18.7	130,832	+2.8
Drunkenness	302,378	−29.2	46,878	−9.9
Disorderly conduct	177,931	−23.0	51,400	−7.3
Vagrancy	10,585	+3.0	2,903	+57.5
All other offenses	1,364,173	+23.8	358,412	+52.3
Suspicion	1,003	−66.1	232	−53.0

SOURCE: Federal Bureau of Investigation. (2002b). 2001 *Uniform Crime Reports* (p. 239). Washington, DC: Author.

Arrest data certainly suggest that the "war on drugs" has translated into a war on women. Between 1992 and 2001, arrests of adult women for drug abuse violations increased by 42.3%, compared to 31.4% for men (FBI, 2002, p. 239). The past decade (1992–2001) has also seen increases in arrests of women for "other assaults" (up 57.7%)—not unlike the pattern seen in girls' arrests. Arrest rates show much the same pattern. In the past decade, arrests of women for drug offenses and other assaults have replaced fraud and disorderly conduct as the most common offenses for which adult women are arrested.

These figures, however, should not be used to support notions of dramatic increases in women's crime. As an example, although the number of adult women arrested between 1992 and 2001 did increase by 15%, that increase followed a slight decline in women's arrests between 1992 and 1993; and by 1997, the overall arrest rate for women has been steadily declining (FBI, 1994, p. 226; 1995, p. 226; 2002, p. 241).

The arrests of women for Part One or Index offenses (murder, rape, aggravated assault, robbery, burglary, larceny theft, motor vehicle theft, and arson) decreased by 9.1% (compared to a decrease in male arrests of 22%) between 1992 and 2001 (FBI, 2002, p. 239; see Table 5.1). Much of this decrease in both women's and men's overall crime index is due to decreases in larceny theft arrests.

Moreover, looking at these offenses differently reveals a picture of stability rather than change over the past decade. Women's share of these arrests (as a proportion of all those arrested for these offenses) rose from 23% to 26% between 1992 and 2001. Women's share of arrests for serious violent offenses rose from 12.4% to 17% during the same period (FBI, 2002, p. 239). This increase is not a result of more murder or robbery arrests, but rather, it is due to aggravated assault. At the other extreme is the pattern found in arrests for prostitution—the only crime among the 29 offense categories tracked by the FBI for which arrests of women account for the majority (63.2%) of all arrests.

Overall, the increase in women's official arrest statistics is largely accounted for by a similar pattern noticed in their juvenile counterparts—increases in other assaults (up 57.7%), drug abuse violations (up by 42.3%), and property offenses, such as check forgery (forgery/counterfeiting, which was up 37.7%) and embezzlement (which was up 78.9%; FBI, 2002, p. 239). Despite an increase in aggravated and other assaults, women's crime is mostly nonviolent in nature—one fifth of all women offenders are arrested for some type of property crime (compared to 8% of men). Here, both property and drug violation arrests are real, because the base numbers are large and, as a result, these offenses make up a large portion of women's official deviance. Whether the increase in other assaults and drug violations arrests (coupled with a consistently large number of arrests for property offenses) are the product of actual changes in women's behavior over the past decade or changes in law enforcement practices is an important question, and one to which we now turn.

HOW COULD SHE? THE NATURE AND CAUSES OF WOMEN'S CRIME

In summary, adult women have been, and continue to be, arrested for minor crimes (generally shoplifting and welfare fraud) and what might be called "deportment" offenses (prostitution, disorderly conduct, and, arguably, "driving under the influence"). Their younger counterparts are arrested for essentially the same crimes, in addition to status offenses (running away from home, incorrigibility, truancy, and other noncriminal offenses for which only minors can be taken into custody). Arrests of adult women, like arrests of girls, have increased for both aggravated and other assaults. Finally, and most important, adult women's arrests for drug offenses have soared.

Where there have been increases in women's arrests for offenses that sound nontraditional, such as embezzlement, careful examination reveals the connections between these offenses and women's place.

Embezzlement

In the case of embezzlement, for which women's arrests increased by 78.9% in the past decade, careful research disputes the notion of women moving firmly into the ranks of big-time, white-collar offenders. Because women are concentrated in low-paying clerical, sales, and service occupations (Renzetti & Curran, 1995), they are "not in a position to steal hundreds of thousands of dollars but they [are] in a position to pocket smaller amounts" (Simon & Landis, 1991, p. 56). Moreover, their motives for such theft often involve family responsibilities rather than a desire for personal gain (Daly, 1989; Zietz, 1981).

Daly's (1989) analysis of gender differences in white-collar crime is particularly useful. In a review of federal "white-collar" crime cases in seven federal districts (which included people convicted of bank embezzlement, income tax fraud, postal fraud, etc.), she found that gender played a substantial role in the differences between men's and women's offenses. For example, of those arrested for bank embezzlement, 60% of the women were tellers and 90% were in some sort of clerical position. By contrast, about half of the men charged with embezzlement held professional and managerial positions (bank officers and financial managers). Therefore, it is no surprise that for each embezzlement offense, men's attempted economic gain was 10 times higher than women's (Daly). In commenting on this pattern, Daly notes, "the women's socio-economic profile, coupled with the nature of their crimes, makes one wonder if 'white collar' aptly described them or their illegalities" (p. 790).

Embezzlement is a particularly interesting offense to "unpack" because it is one of the offenses for which, if present trends continue, women may comprise about half of those charged (Renzetti & Curran, 1995, p. 310). In fact, women composed about half of those charged with embezzlement in 2001 (FBI, 2002, p. 239). Yet these increases cannot be laid at the door of women breaking into traditionally "male" offense patterns. Women's increased share of arrests for embezzlement is probably an artifact of their presence in low-level positions that make them more vulnerable to frequent checking and hence more vulnerable to detection (Steffensmeier & Allan, 1995). Combining this with these women's lack of access to resources to "cover" their thefts prompts Steffensmeier and Allan to draw a parallel between modern women's involvement in embezzlement and increases in thefts by women in domestic service a century ago.

Driving Under the Influence

Arrests of women driving under the influence (DUI) account for almost 1 arrest in 10 of women (FBI, 2002, p. 239). One study (Wells-Parker, Pang, Anderson, McMillen, & Miller, 1991) found that women arrested for DUI tended to be older than men (with nearly half of the men but less than a third of the women under 30), more likely to be "alone, divorced or separated," and to have fewer serious drinking problems and fewer extensive prior arrests for DUI or "public drunkenness" (Wells-Parker et al., 1991, p. 144). Historically, women were arrested for DUI only if "the DUI involved a traffic accident or physical/verbal abuse of a police officer" (Coles, 1991, p. 5). These patterns have probably eroded in recent years because of public outrage over drinking and driving and an increased use of roadblocks. Changes in police practices, rather than changes in women's drinking, could easily explain the prominence of this offense in women's official crime patterns.

Women tend to drink alone and deny treatment (Coles, 1991). Also, in contrast to men, they tend to drink for "escapism and psychological comfort" (Wells-Parker et al., 1991, p. 146). For these reasons, intervention programs that attempt to force women to examine their lives and the quality of relationships, which tend to work for male DUI offenders, are not successful with women. Indeed, these interventions could "exacerbate a sense of distress, helplessness and hopelessness" that could, in turn, trigger more drinking (Wells-Parker et al., p. 146).

Larceny Theft/Shoplifting

Women's arrests for larceny theft are composed largely of arrests for shoplifting. Steffensmeier (1980) estimates that perhaps as many as four fifths of all arrests on larceny charges are for shoplifting. Cameron's (1953) early study of shoplifting in Chicago explains that women's prominence among those arrested for shoplifting may not reflect greater female involvement in the offense but rather differences in the ways men and women shoplift. Her research revealed that women tend to steal more items than men, to steal items from several stores, and to steal items of lesser value. Store detectives explained this pattern by saying that people tended to "steal the same way they buy" (Cameron, p. 159). Men came to the store with one item in mind. They saw it, took it, and left the store. Women, on the other hand, shopped around. Because the chance of being arrested increases with each item stolen, Cameron felt that the stores probably underestimated the level of men's shoplifting.

Although women stole more items than men, the median value of adult male theft was significantly higher than that of women's (Cameron, 1953, p. 62). In addition, more men than women were defined as "commercial shoplifters" (people who stole merchandise for possible resale).

Perhaps as a result of women's shopping and shoplifting patterns, studies done later (Lindquist, 1988) have found that women constitute 58% of those caught shoplifting. Steffensmeier and Allan (1995) go so far as to suggest that shoplifting may be regarded as a prototypically female offense. Shopping is, after all, part of women's "second shift" of household management, housework, and child care responsibilities (Hochschild, 1989). Shoplifting can be seen as a criminal extension of expected and familiar women's work.

Even the reasons for shoplifting are gendered. Men, particularly young men, tend to view stealing as part of a broader pattern of masculine display of "badness" and steal items that are of no particular use to them (Steffensmeier & Allan, 1995). At the other extreme, they may be professional thieves and thus more likely to escape detection (Cameron, 1953).

Girls and women, on the other hand, tend to steal items that they either need or feel they need but cannot afford. As a result, they tend to steal from stores and to take things such as clothing, cosmetics, and jewelry. Campbell (1981) notes that women—young and old—are the targets of enormously expensive advertising campaigns for a vast array of personal products. These messages, coupled with the temptations implicit in long hours spent "shopping," lead to many arrests of women for these offenses.

Despite some contentions that women actually shoplift more than men, self-report data in fact show few gender differences in the prevalence of the behavior (see Chesney-Lind & Shelden, 1992, for a review of these studies). What appears to be happening is that girls and women shoplift in different ways than men. In addition, they are more often apprehended because store detectives expect women to shoplift more than men and thus watch women more closely (Morris, 1987).

BIG TIME/SMALL TIME

English (1993) approached the issue of women's crime by analyzing detailed self-report surveys she administered to a sample of 128 female and 872 male inmates in Colorado. She examined both "participation rates" and "crime frequency" figures for a wide array of different offenses. She found few differences in the participation rates of men and women, with the exception of three property crimes. Men were more likely than women to report participation in burglary, whereas women were more likely than men to have participated in theft and forgery. Exploring these differences further, she found that women "lack the specific knowledge needed to carry out a burglary" (p. 366).

Women were far more likely than men to be involved in "forgery" (it was the most common crime for women and fifth out of eight for men). Follow-up research on a subsample of "high crime"-rate female respondents revealed that many had worked in retail establishments and therefore "knew how much time they had" between stealing the checks or credit cards and having them reported (English, 1993, p. 370). The women said that they would target "strip malls" where credit cards and bank checks could be stolen easily and used in nearby retail establishments. The women reported that their high-frequency theft was motivated by a "big haul," which meant a purse with several hundred dollars in it, in addition to cards and checks. English concludes that "women's over representation in low-paying, low status jobs" increases their involvement in these property crimes (p. 171).

English's (1993) findings with regard to two other offenses, for which gender differences did not appear in participation rates, are worth exploring here. She found no difference in the "participation rates" of women and men in drug sales and assault. However, when examining the frequency data, English found that women in prison reported making significantly more drug sales than men but not because they were engaged in big-time drug selling. Instead, their high number of drug sales occurred because women's drug sales were "concentrated in the small trades (i.e., transactions of less than $10)" (p. 372). Because they made so little money, English found that 20% of the active women dealers reported making 20 or more drug deals per day (p. 372).

A reverse of the same pattern was found when English (1993) examined women's participation in assault. Here, slightly more (27.8%) women than men (23.4%) reported assaulting someone in the past year. However, most of these women reported making only one assault during the study period (65.4%), compared to only about a third of the men (37.5%).

English (1993) found that "economic disadvantage" played a role in both women's and men's criminal careers. Beyond this, however, gender played an important role in shaping women's and men's response to poverty. Specifically, women's criminal careers reflect "gender differences in legitimate and illegitimate opportunity structures, in personal networks, and in family obligations" (p. 374).

PATHWAYS TO WOMEN'S CRIME

As with girls, the links between adult women's victimization and crimes are increasingly clear. As was noted in earlier chapters, the backgrounds of adult women offenders hint at links between childhood victimization and adult offending. Experiencing gender and racial oppression, those groups of women who are most socially marginalized are particularly vulnerable to both problems—abuse/victimization and involvement in illegal activity (Richie, 2000). For example, a 1996 survey of women in prison reported that at least half of them experienced sexual abuse before their incarceration—a much higher rate than what is reported in the general population (Richie, p. 5). Other studies have documented the link between women's experiences with physical and/or sexual violence and their involvement with illegal drugs (Harlow, 1999).

Widom (2000) demonstrates in her work the importance of understanding women's experiences of abuse and neglect during childhood and their entrance into criminal activity. Abused and neglected girls are nearly twice as likely to be arrested as juveniles, twice as likely to be arrested as adults, and 2.4 times more likely to be arrested for violent crimes (p. 29). They are "more likely to use alcohol and other drugs and turn to criminal and violent behaviors when coping with stressful life events" (p. 33). Widom explains (and as previously mentioned in Chapter 2) victimization prompts girls' entry into delinquency as they try and flee their abusive environments. With deficits in cognitive abilities and achievement and few positive relationships or social controls, these girls end up on the streets with hardly any legitimate survival skills (p. 30). Their experiences with victimization and violence may result in lowered self-esteem, a lack of sense of control over one's life, and behavioral inclinations for crime and violence. Consequently, they become women with few social or psychological resources for successful adult development.

Family problems and violence—such as death, disruption, abuse and neglect, poverty, and drug/alcohol addiction—produce gendered effects for boys and girls. Although both boys and girls who grow up in family environments riddled with crime and violence have higher propensities to model such behavior (Widom, 2000), girls must also negotiate gender oppression. Such oppression confines girls and women to a dichotomous characterization: weak and dependent as well as sexually uncontrollable (Gelsthorpe, 1989; Girschick, 1999). Girschick (p. 30), in her stories of women in prison, points out an important consequence: Abused females are the least likely to have been affected in a positive way by the challenge of the women's movement to traditional gender roles and expectations. In a desperate need to have someone close to them, they often feel powerless, have limited options for change, and meet with continued abuse and violence. These women are trapped by patriarchy (and for women of color, racial and ethnic discrimination), their gender identity, their loyalty to their partners, and the violence itself (Girschick, p. 58; Richie, 1996).

Other studies have also demonstrated this important link between childhood trauma and adult criminality. Gilfus (1992) interviewed 20 incarcerated women and documented how such childhood injuries were linked to adult crimes in women. Gilfus extends the work of Miller (1986) and Chesney-Lind and Rodriguez (1983) on the ways in which women's backgrounds color their childhoods and ultimately their adulthoods. She conducted in-depth interviews in 1985 and 1986 with the women in a Northeastern women's facility that, at the time, served as both a jail and a prison. From these lengthy interviews, she reconstructed "life event histories" for each of the women. The group had a mean age of 30 (ranging from 20 to 41 years of age), and included 8 African American and 12 white women. All of the women had life histories of what Gilfus characterized as "street crimes"—by which she meant prostitution, shoplifting, check or credit card fraud, and drug law violations. Their current offenses included assault and battery; accessory to rape; breaking and entering; and multiple charges of drug possession, larceny, and prostitution (Gilfus, p. 68). Sentence lengths ranged, for this group, from 3 months to 20 years.

Most of the women were single mothers, three-quarters were intravenous drug users, and almost all (17) had histories of prostitution (7 of the women had begun as teenage prostitutes). Most of these women had grown up with violence; 13 of the 20 reported childhood sexual abuse, and 15 had experienced "severe child abuse" (Gilfus, 1992, p. 70). There were no differences in the levels of abuse reported by black and white respondents, although African American women grew up in families that were more economically marginalized than their white counterparts. Although some women's childhood memories were totally colored by their sexual abuse, for most of the women in Gilfus's sample, coping with and surviving multiple victimization was the more normal pattern. In the words of one of these women, "I just got hit a lot. . . . 'Cause they would both drink and they wouldn't know the difference. Mmm, picked up, thrown against walls, everything, you name it" (p. 72).

Despite the abuse and violence, these women recall spending time trying to care for and protect others, especially younger siblings, and attempting to do housework and even care for their abusive and drug or alcohol dependent parents. They also recall teachers who ignored signs of abuse and who, in the case of African American girls, were hostile and racist. Ultimately, 16 out of the 20 dropped out of high school (Gilfus, 1992, p. 69). The failure of the schools to be responsive to these young women's problems meant that the girls could perceive no particular future for themselves. Given the violence in their lives, drugs provided these girls with a solace found nowhere else.

Many (13) ran away from home as girls. "Rape, assault, and even attempted murder" were reported by 16 of the 20, with an average of three "rape or violent rape attempts" per woman; many of these occurred in the context of prostitution but when the women attempted to report the assault, the police simply "ridiculed" the women or threatened to arrest them. In some cases, the police would demand sexual services for not arresting the woman (Gilfus, 1992, p. 79).

Violence also characterized these women's relationships with adult men; 15 of the 20 had lived with violent men. The women were expected to bring in money, generally through prostitution and shoplifting. These men functioned as the women's pimps but also sold drugs, committed robberies, or fenced the goods shoplifted by the women. Thirteen of the women had become pregnant as girls but only four kept their first baby. Most of the women had subsequent children whom they attempted to mother despite their worsening addictions, and they tended to rely on their mothers (not their boyfriends) to take care of their children while in prison. The women continued to see their criminal roles as forms of caretaking, taking care of their children and of their abusive boyfriends. As Gilfus (1992) puts it, "the women in this study consider their illegal activities to be form of work which is undertaken primarily from economic necessity to support partners, children, and addictions" (p. 86). Gilfus further speculates that violence "may socialize women to adopt a tenacious commitment to caring for anyone who promises love, material success, and acceptance" (p. 86), which, in turn, places them at risk for further exploitation and abuse.

The interviews Arnold (1995) conducted, based on this same hypothesis, with 50 African American women serving sentences in a city jail and 10 additional interviews with African American women in prison are an important addition to the work of Gilfus (1992). Arnold notes that African American girls are not only sexually victimized but are also the victims of "class oppression." Specifically, she notes that "to be young, Black, poor and female is to be in a high-risk category for victimization and stigmatization on many levels" (p. 139). Growing up in extreme poverty means that African

American girls may turn earlier to deviant behavior, particularly stealing, to help themselves and their families. One young woman told Arnold that "my father beat my mother and neglected his children. . . . I began stealing when I was 12. I hustled to help feed and clothes the other [12] kids and help pay the rent" (p. 139).

Thus, the caretaking role noted in women's pathways to crime is accentuated in African American families because of extreme poverty. Arnold (1995) also noted that economic need interfered with young black girls' ability to concentrate on schoolwork and attend school. Finally, Arnold (p. 140) noted, as had Gilfus (1992), that African American girls were "victimized" by the school system; one of her respondents said that "some [of the teachers] were prejudiced, and one had the nerve to tell the whole class he didn't like black people" (p. 140). Most of her respondents said that even if they went to school every day, they did not learn anything. Finally, despite their desperate desire to "hold on . . . to conventional roles in society," the girls were ultimately pushed out of these, onto the streets, and into petty crime (p. 141).

The mechanics of surviving parental abuse and educational neglect were particularly hard on young African American girls, forcing them to drop out of school, on to the streets, and into permanent "structural dislocation" (Arnold, 1995, p. 143). Having no marketable skills and little education, many resorted to "prostitution and stealing" while further immersing themselves in drug addiction.

BEYOND THE STREET WOMAN: RESURRECTING THE LIBERATED FEMALE CROOK?

Issues of women's violence and the relationship between that violence and other changes in women's environment are recurring themes in discussions of women's crime. As we saw in Chapter 3, a persistent theme in women's criminality is the presumed link between efforts to improve women's economic and political position and levels of girls' and women's crime—particularly violent crime. We also saw that there is nothing particularly "new" about the concern. During the early 1970s, newspapers and periodicals were full of stories about the "new female criminal" (Foley, 1974; Klemesrud, 1978; Los Angeles Times Service, 1975; Nelson, 1977; Roberts, 1971). Presumably inspired by the women's movement, the female criminal supposedly sought equality in the underworld just as her more conventional counterparts pursued their rights in more acceptable arenas.

Such media accounts, like contemporary "girlz in the hood" stories, generally relied on two types of evidence to support the alleged relationship between the women's rights movement and increasing female criminality: FBI statistics showing dramatic increases in the number of women arrested for nontraditional crimes, and sensationalistic accounts of women's violence. In the 1970s, the activities of female political activists such as Leila Khaled, Bernardine Dohrn, and Susan Saxe were featured. Of course, women's involvement in political or terrorist activity is nothing new, as the activities of Joan of Arc and Charlotte Corday demonstrate.

Arrest data collected by the FBI seem to provide more objective evidence that dramatic changes in the number of women arrested were occurring during the years associated with the second wave of feminist activity. For example, between 1960 and 1975, arrests of adult women went up 60.2% and arrests of juvenile women increased a startling 253.9%. In specific, nontraditional crimes, the increases were even more astounding. For example, between 1960 and 1975, the number of women arrested for murder was up 105.7%, forcible rape arrests increased by 633.3%, and robbery arrests were up 380.5% (FBI, 1973, p. 124; 1976, p. 191).

Law enforcement officials were among the earliest to link these changes to the movement for female equality. "The women's liberation movement has triggered a crime wave like the world has never seen before," claimed Chief Ed Davis of the Los Angeles Police Department (Weis, 1976, p. 17). On another occasion, he expanded on his thesis by explaining that the "breakdown of motherhood" signaled by the women's movement could lead to "the use of dope, stealing, thieving and killing" (Los Angeles Times Service, 1975, p. B4). Other officials, such as Sheriff Peter Pitchess of California, made less inflammatory comments that echoed the same general theme: "As women emerge from their traditional roles as housewife and mother, entering the political and business fields previously dominated by males, there is no reason to believe that women will not also approach equality with men in the criminal activity field" (Roberts, 1971, p. 72).

Law enforcement officials were not alone in holding this position; academics like Adler (1975a, 1975b) also linked increases in the number of women arrested to women's struggle for social and economic equality. Adler noted, for example, that in the middle of the twentieth century, we are witnessing the simultaneous rise and fall of women. Rosie the Riveter of World War II has become Robin the Rioter or Rhoda the Robber of the Vietnam era. Women have lost more than their chains. For better or worse, they have lost many of the restraints which kept them within the law. (1975b, p. 24)

Such arguments, it turns out, are nothing new. The first wave of feminism also saw an attempt to link women's rights with women's crime. Smart (1976), for example, found the following comment by W. I. Thomas written in 1921, the year after the ratification of the 19th Amendment guaranteeing women the right to vote:

The modern age of girls and young men is intensely immoral, and immoral seemingly without the pressure of circumstances. At whose door we may lay the fault, we cannot tell. Is it the result of what we call "the emancipation of woman," with its concomitant freedom from chaperonage, increased intimacy between the sexes in adolescence, and a more tolerant viewpoint towards all things unclean in life? (Smart, 1976, pp. 70–71)

Students of women's crime were also quick to note that the interest in the female criminal after so many years of invisibility was ironic and questioned why the new visibility was associated with "an image of a woman with a gun in hand" (Chapman, 1980, p. 68). Chapman concluded that this attention to the female criminal was "doubly ironic" because closer assessments of the trends in women's violence did not support what might be called the "liberation hypothesis" and, more to the point, data showed that the position of women in the mainstream economy during those years was "actually worsening" (pp. 68–69).

To be specific, although what might be called the "liberation hypothesis" or "emancipation hypothesis" met with wide public acceptance, careful analyses of changes in women's arrest rates provided little support for the notion. Using national arrest data supplied by the FBI and more localized police and court statistics, Steffensmeier (1980, p. 58) examined the pattern of female criminal behavior for the years 1965 through 1977 (the years most heavily affected by the second wave of feminist activity). By weighting the arrest data for changes in population and comparing increases in women's arrests to increases in men's arrests, Steffensmeier concluded that "females are not catching up with males in the commission of violent, masculine, male-dominated, serious crimes (except larceny) or in white collar crimes" (p. 72). He did note increases in women's arrests in the Uniform Crime Report categories of larceny, fraud, forgery, and vagrancy, but by examining these increases more carefully, he demonstrated that they were due almost totally to increases in traditionally female crimes, such as shoplifting, prostitution, and passing bad checks (fraud).

Moreover, Steffensmeier (1980) noted that forces other than changes in female behavior were probably responsible for shifts in the numbers of adult women arrested for these traditionally female crimes. The increased willingness of stores to prosecute shoplifters, the widespread abuse of vagrancy statutes to arrest prostitutes combined with a declining use of this same arrest category to control public drunkenness, and the growing concern with "welfare fraud" were all social factors that he felt might explain changes in women's arrests without necessarily changing the numbers of women involved in these activities.

Steffensmeier's (1980) findings confirm the reservations that had been voiced earlier by Simon (1975), Rans (1975), and others about making generalizations solely from dramatic percentage increases in the number of women's arrests. These reservations are further justified by current arrest data that suggest that the sensationalistic increases of the early 1970s were not indicative of a new trend. Between 1976 and 1979, for example, the arrests of all women rose only 7.1%, only slightly higher than the male increase of 5.8% for the same period (Chesney-Lind, 1986).

Finally, as previously mentioned, women offenders of the 1970s were unlikely targets for the messages of the largely middle-class women's movement. Women offenders tend to be poor, members of minority groups, with truncated educations and spotty employment histories. These were precisely the women whose lives were largely unaffected by the gains, such as they were, of the then white, middle-class women's rights movement (Chapman, 1980; Crites, 1976). Crites, for example, noted that "these women rather than being recipients of expanded rights and opportunities gained by the women's movement, are, instead, witnessing declining survival options" (1976, p. 37). Research on the orientations of women offenders to the arguments of the women's movement also indicated that, if anything, these women held very traditional attitudes about gender (see Chesney-Lind & Rodriguez, 1983, for a summary of these studies).

To summarize, careful work on the arrest trends of the 1970s provided no support for the popular liberation hypothesis of women's crime. Changes in women's arrest trends, it turned out, better fit arguments of woman's economic marginalization than of her liberation (Simon & Landis, 1991).

THE REVIVAL OF THE
"VIOLENT FEMALE OFFENDER"

The failure of careful research to support notions of radical shifts in the character of women's crime went almost completely unnoticed in the popular press. As a result, there was apparently nothing to prevent a recycling of a revised "liberation" hypothesis a decade and a half later.

As noted earlier, one of the first articles to use this recycled hypothesis was a 1990 article in the *Wall Street Journal*, titled "You've Come a Long Way, Moll," that focused on increases in the number of women arrested for violent crimes. This article opened with a discussion of women in the military, noting that "the armed forces already are substantially integrated" and moved from this point to observe that "we needn't look to the dramatic example of battle for proof that violence is no longer a male domain. Women are now being arrested for violent crimes—such as robbery and aggravated assault—at a higher rate than ever before recorded in the U.S." (Crittenden, 1990, p. A14).

As noted in Chapter 3, many of the articles to follow dealt with young women (particularly girls in gangs), but there were some exceptions. For example, the "Hand That Rocks the Cradle Is Taking Up Violent Crime" (Kahler, 1992) focused on increases in women's imprisonment and linked this pattern to "the growing number of women committing violent crimes" (p. 3A). And in April 2000, in the *New York Times'* series on "Rampage Killers," the author was quick to point out by the third sentence of the story that rampage killers are "mostly white men, but a surprising number are women" (Fessenden, 2000), when in fact, it is even rarer for women to commit multiple victim homicides than single victim ones (about 4% of all such crimes).

Comparing the arrest rates that prompted the first media surge of reporting on the "liberation" hypothesis with the arrest rates from the second wave of media interest, it is evident that little has changed. As noted earlier in this chapter, women's share of violent crime has remained more or less stable, though arrests of women for "other assaults" did climb by 57.7% in the past decade (compared to 8.7% for men; FBI, 2002). In the past 5 years, however, women's arrests for other assaults climbed only 7.5%, while all other categories of violent crime arrests, including weapons carrying, decreased for women (p. 241).

The news media were not alone in their interest in women's violence—particularly violent street crime. In a series of articles (Baskin, Sommers, &

Fagan, 1993; Sommers & Baskin, 1992, 1993), the authors explore the extent and character of women's violent crime in New York. Prompted by an account in their neighborhood paper of two women who shot another woman in a robbery, the authors note that their research "has led us to the conclusion that women in New York City are becoming more and more likely to involve themselves in violent street crime" (Baskin et al., 1993, p. 401). Some of the findings that brought them to this conclusion follow.

In one study, Sommers and Baskin (1992) used arrest data from New York City (and arrest histories of 266 women) to argue that "black and Hispanic females exhibited high rates of offending relative to white females." They further argue that "violent offending rates of black females parallel that of white males" (p. 191). Included in their definition of "violent" crimes is murder, robbery, aggravated assault, and burglary (apparently classified as a violent crime in New York City, but classified as a property crime by the FBI).

The authors explain that this pattern is a product of "the effects of the social and institutional transformation of the inner city" (Sommers & Baskin, 1992, p. 198). Specifically, the authors contend that "violence and drug involvement" are adaptive strategies in underclass communities that are racked by poverty and unemployment. Both men and women, they argue, move to crime as a way of coping with "extreme social and economic deprivation" (p. 198).

A second study (Sommers & Baskin, 1993) further explores women's violent offenses by analyzing interview data from 23 women arrested for a violent felony offense (robbery or assault) and 65 women incarcerated for such an offense. Finding a high correlation between substance abuse and the rate of violent crime (particularly for those who committed robbery and robbery with assault), they also noted that "the women in our study who were involved in robbery were not crime specialists but also had a history of engagement in nonviolent theft, fraud, forgery, prostitution, and drug dealing" (p. 142). In fact, they comment that "these women are not roaming willy-nilly through the streets engaging in 'unprovoked' violence" (p. 154).

Just how involved these women were in more traditional forms of female crime was not apparent in the text of this article. In an appendix, however, the role played by a history of prostitution in the lives of these women offenders is particularly clear. As an example, the women who reported committing both robbery and assault also had the highest rate of involvement with prostitution (77%; Baskin & Sommers, 1993, p. 159).

In a third paper, Baskin et al. (1993) explore "the political economy of street crime." This work, which appears to be based on the New York arrest data and a discussion of the explosion of crack selling in the city, explores the question, "Why do black females exhibit such relatively high rates of violence?" (p. 405). Convinced that the concentration of poverty is associated positively with the level of criminal activity, regardless of race, the authors then conclude that "the growing drug markets and a marked disappearance of males" combine with other factors in underclass communities "to create social and economic opportunity structures open to women's increasing participation in violent crime" (p. 406).

The authors further suggest that traditional theories of women's offenses, particularly those that emphasize gender and victimization, do not adequately explain women's violent crime. Their work, they contend, "confirms our initial sense that women in inner city neighborhoods are being pulled toward violent street crime by the same forces that have been found to affect their male counterparts (e.g., peers, opportunity structures, neighborhood effects)" (Baskin et al., 1993, p. 412). They conclude that the socioeconomic situation in the inner city, specifically as it is affected by the drug trade, creates "new dynamics of crime where gender is a far less salient factor" (p. 417).

These authors argue that in economically devastated inner cities such as New York, women's violence—particularly the violence of the women of color—does not need to be considered in terms of the place of these women in patriarchal society (e.g., the effect of gender in their lives). Instead, they contend that these women (like their male counterparts) are being drawn to violence and other forms of traditionally male crimes for the same reasons as men.

This turns the "liberation" hypothesis on its head. Now, it is not presumed economic gain that promoted "equality" in crime, but rather it is economic marginalization that causes women to move out of their "traditional" roles into the role of criminal. Is that really what is going on?

This chapter has already cast doubt on the notion that there has been any dramatic shift in women's share of violent crime (at least as measured by arrest statistics). This chapter has also provided evidence that women's participation in offenses that sound "nontraditional" (such as embezzlement, DUI, and larceny theft) is deeply affected by the "place" of women in society. Both of these findings cast doubt on claims for the existence of a new, violent street criminal class of women—at least without first providing a more detailed exploration of trends in women's involvement in crimes such as robbery and drug selling. Because these offenses, particularly drug use and sale, feature so prominently in the debate about the nature of adult women's offenses, they are explored in detail in the next chapter.

Has there been an increase in women's participation in traditionally male types of crime, such as violent crime? Does women's search for "equality" with men have a darker side, as suggested by some of the arrest statistics reviewed in this chapter? To answer this question fully, the next chapter explores the current context of women's violence. Specifically, the chapter explores the relationship between apparently male offenses (such as robbery and drug selling) and more traditional female offenses (such as prostitution).

DRUGS, VIOLENCE,
AND WOMEN'S CRIME

With Karen Joe Laidler

———•◦•———

T he past few years have seen a rebirth of interest in the "new" female offender and, as was the case with girls, this renewed interest is not good news for women on the margins. As shown in the last chapter, the media and some scholars have used the increase in women's arrests for crimes of violence and the increase in women's involvement in drug use and sales as evidence of a fundamental change in the nature of women's crime.

This perspective contends that the new female offender is not only gaining an egalitarian footing with her male counterpart in crime generally, but in drug use and sales specifically. Women's involvement in drug use, the argument continues, leads to women's involvement in robbery and other forms of nontraditional offenses. This new version of the liberation or emancipation hypothesis draws heavily on current thinking about the "underclass" (see Baskin et al., 1993). These researchers contend that it is not race or gender but one's place in the economic order (social class) that drives life in the deindustrialized ghetto neighborhoods of the 1990s (Sommers, Baskin, & Fagan, 2000). According to this argument, the deindustrialization of the city has contributed to degendering male and female relationships and has made women more capable of violence and other forms of traditionally masculine crime.

This chapter takes issue with this construction of women's lives and women's crime in a number of ways. First, the chapter draws from an ethnographic study to focus on the role of drugs in women's lives in a multiethnic community. Then, the chapter moves to a more quantitative discussion of the role of robbery in women's offenses in that same community. Finally, what is known about other forms of women's violence is reviewed.

These discussions provide direct challenges to the notion that gender is unimportant in the lives of women of color in neighborhoods devastated by poverty. Instead, these accounts demonstrate that, far from being liberated by economic dislocation, these women's lives are deeply affected by the sex/gender system,[1] which continues to play a major role in the lives of all those involved in violence and drug use. They also illuminate the ways in which race and class shape the choices that girls and women make, and forcefully document the ways in which girls' problems are linked to their involvement as adult women in crime.

DRUG USE IN A MULTIETHNIC COMMUNITY

This portion of the chapter draws from an ethnography (Joe, 1995a, 1995b, 1996a, 1996b; Joe & Morgan, 1997; Morgan, Beck, Joe, McDonnell, & Gutierrez, 1994; Morgan & Joe, 1996, 1997) exploring the social world of a group of women in Hawaii who are moderate to heavy users of methamphetamine or crack cocaine.[2] The study explores the ways in which these women define and experience violence from early childhood through their drug use careers, and the strategies and resources they draw upon to cope with and resist violence. Central to both the problems these women experience and the solutions they craft are their cultural roots, particularly as expressed in extended family networks.

Hawaii has been described as "futuristic," because it was the first state to include a vast array of ethnic groups beyond the familiar black/white divide that once characterized the discussion of race in America (Hippensteele & Chesney-Lind, 1995). Less, though, is known about the cultural groups in the islands, so a brief discussion of their situation and worldview is important.

Despite the unique character of the many different Asian ethnic cultures, Asian American (Japanese American, Chinese American, and Filipino American) families generally have been characterized in relation to the "old

world" customs—patriarchy, filial piety and respect, "saving face," and preserving the honor and reputation of the family. These "ideal" traits of Asian American families are assumed to provide cohesiveness and harmony and, simultaneously, protect Asian Americans from experiencing social and economic problems, such as delinquency, illicit drug use, and poverty.

This "model minority" stereotype is exactly that—a stereotype. There is, in fact, great diversity among the more than 32 Asian American ethnic groups with different histories, experiences, and cultures. And, as shall be seen, this stereotype is far from the reality experienced by Asian American women drug users.

Pacific Islander cultures (Native Hawaiian, part-Hawaiian, and Samoan) feature a strong focus on family and kinship, and strong models of female leaders (Linnekin, 1990; Nunes & Whitney, 1994). However, their communities, like other Native American groups, are struggling with the profound effect of centuries of colonialism and, particularly in Hawaii, poverty. Others (particularly Filipino and Samoan women) are also experiencing the effect of recent immigration on family life, with all its attendant strains and dislocations. Their experiences contrast with other Asian ethnic groups, particularly the Japanese and Chinese, whose families immigrated to Hawaii during the early part of the 20th century to work in the plantations and have since moved into the mainstream of the economic life of the islands. In essence, the islands are both ethnically and economically stratified, and groups such as the Native Hawaiians, Filipinos, Samoans, and other recent immigrant groups tend to have the lowest incomes (Okamura, 1990).

There is considerable diversity among the women in these groups. However, one thread that remains constant in the lives of virtually all of these women is poverty and its effect on their ethnic communities. As the women's accounts in this chapter will show, their families constantly struggled on the economic margins, and this became a significant source of their stress and problems with illicit drugs. As we let them tell their life histories, we are able to see firsthand how gender, family, and class actually "work" in the lives of these young women.

By looking carefully at the lives of these individual girls and women and listening to their voices, we can begin to understand how the families in these communities cope with economic marginality. We will see that this experience can affect family relationships in contradictory ways. Specifically, the family can act as both a protector and as a facilitator of problems such as drug abuse (Joe 1995b, 1996b). And, as these accounts will document, these women's lives underscore the paradoxical nature of the cultural claims and strong ties associated with many minority families. Although they experienced conflict and violence at home, their families were their most significant resource for support.

A PROFILE OF THE WOMEN

Although Asian and Pacific Americans are among the fastest growing ethnic groups, they and their problems, including illicit drug use, remain almost completely invisible on the national landscape.[3] Yet as the 1992 Los Angeles riots and the ethnic complexity of the issues that surfaced during those events have taught us, race relations (and problems with racism) have moved past the narrow black/white paradigm. As the country watched, African Americans, Hispanic Americans, and Korean Americans poured onto the streets and into the collective consciousness of the country. In that moment, the general public was made forcibly aware that the United States has moved into a world of multiple cultures and multiple ethnicities. Listening to the women included in this chapter, then, will shed much-needed light on the problems of groups that have traditionally been coded as "other" and ignored by traditional criminological scholarship.

The ethnic heritage of the women interviewed reflects Hawaii's diversity and complexity and its historically high out-marriage rate.[4] Over half of the women were part-Hawaiian. Nearly a third identified as Filipina, with many indicating they were of mixed ethnicity. The remainder of the group were Samoan, mixed Portuguese, Chinese, and Japanese. With the exception of two Filipinas, all of the women were born in Hawaii and grew up in lower- and working-class families. The majority of the women were in their mid to late 20s, had never been married, and had at least one child. Few had gone beyond high school, and nearly a third had dropped out prior to completing 12th grade. The majority of the women in the sample were unemployed, lived below the poverty line, and survived on government assistance, their families, or illegal activities.

THE FAMILY: CONFLICT AND COMFORT

In many Asian American and Pacific Islander cultures, the family is based on an extended kinship network. Cultural differences between families may be found in the nature of the interaction, extent of the network, gender roles, and expectations. These are further affected by economic and social change. In Hawaii, a distinctive extended kinship network has emerged and is loosely known as the ohana system.[5] Although ohana originates from the Hawaiian familial tradition of solidarity, shared involvement, and interdependence, today it reflects the blending of Hawaii's many Asian Pacific American ethnic groups and their cultures, which have come to be conceptualized as "local" (Chinen, 1994; Okamura, 1994; Yamamoto, 1979). In its contemporary form, ohana maintains the familial spirit and traditions of cooperation and unity. However, those persons who are not blood-related but integral to the family are often included in the kinship network.

Most of the women in this study grew up, to differing degrees, in an extended family system. Given the extremely high cost of living in Hawaii, this kinship network provided their economically strained families with a stable source of support. The women described moving from the households of their parents to other members of the family network (aunts, uncles, grannies, cousins) at various times in their life. Yet the ohana system serves not only as an immediate resource to cope with families' financial problems but also as a relief mechanism for the heated tension that often emerges from economic marginality. Take a look at the following description of the life of one of the women in the study, Mary:

> Mary is a 23 year old Chinese Hawaiian woman who is the fourth child of six. Her mother has been married three times, and her children are from different marriages. While Mary's older sister was sent to stay with their "rich" aunt, she and her younger brother lived with their grandmother during their early childhood years. While Mary believes that she was sent to live with her grand- mother to "take care of her and to help her with the house and cook," in light of her young age, it is more likely that her mother relied on her own mother to help care for two of her children while she worked and looked after the other children and a physically abusive husband who suffered from severe diabetes. During her teen years, Mary returned to her mother's home and helped care for her diabetic stepfather. (Study Interview #446)

The pressures associated with economic marginality may also manifest themselves in parental alcohol and other drug use. Although nearly half of the women's parents used marijuana and over a third of them used cocaine, alco- hol appeared to be the most problematic. Forty percent of the women report that their parents had problems with alcohol. However, this figure is probably an underestimation because many tried to rationalize their parents' alcohol use. Again, consider the life of Joanne (a 44-year-old homeless woman):

> Joanne states that her father consumed several cases of beer on the weekends, but was only a "recreational drinker" because he "never missed work due to his drinking" and, most important, provided for his family. She had her first drink at 22 years of age when her father became seriously ill and died, and, "for the next ten years stayed in an unconscious drunken state by noontime everyday." (Study Interview #551)

Parental alcohol or drug use was often connected with physical or sexual violence. In some cases, the violence was severe and the extended family system was unable to provide a long-term sanctuary. For Susan, a 19-year-old woman, the domestic violence became horrific:

> Susan remembers from about the age of five that her father would routinely beat up her mother to the point where she would be unable to walk. Subsequently her father would come looking for her or her mother would take out her own anger and hostility by beating on Susan and her siblings. Both parents were heavily involved in drugs, and her father was a dealer. She describes having a loose family structure as her father had several children by other women.
>
> While growing up, she was exposed to many "adult" situations including drug deals and hanging out in bars. Her father was sent to prison for hanging a man on a fence and beating him to death while drunk. At 14, an unknown teenage male raped her at a family function. She tried to isolate herself, but when her mother learned of the incident, she punished her for "promiscuity" by repeatedly hitting her on the head and sending her to a group home for troublesome teenagers. (Study Interview #462)

Susan's case also illustrates how the intensity of family pressures heightens as these young women are expected to conform to culturally proscribed gender roles and expectations.[6] Although she had been raped, her family believed that it was she, not her assailant, who was at fault. Such a situation is doubly ironic because traditional Hawaiian gender norms would not have condemned her (Linnekin, 1990). However, extensive contact with the West has meant that many contemporary Hawaiian families participate in the sexual double standard that, as we saw earlier, decrees that women must guard their sexual "reputation" and are at fault for all sexual behavior, even rape.

Susan's life also illustrates a process called "parentalization." Like many girls in economically marginalized households, these women—as children—assumed adult responsibilities because both parents worked long hours or one parent had left the family (either temporarily or permanently). Many described having to become the "parent" of the house, caring for their younger siblings and managing the domestic chores of the house.

The strain of poverty, combined with cultural expectations about "being a good girl" and heavy parental alcohol consumption, exacerbates family violence. Helen, a 38-year-old woman, recalls her girlhood:

> I come from a family of six children and I'm the fourth. We are all scattered. One brother is in prison and one passed away. When we was growing up we lived with both my parents. They stayed married until my dad passed away. Home was very strict. My dad was an alcoholic so he couldn't hold a job. He always had a strict hand on us. Discipline kind. He was either drunk or coming down from a hangover when he hit us. My mom was the one that went to work. Beatings were all the time from my dad. Severe kind with belt buckles.
>
> The last time my dad hit me was when I was 17 years old. He found out that I was smoking cigarettes. I was almost 18. My youngest brother was able to drink with him, smoke cigarettes, and pot with him! But not me. The boys could do what they wanted. My mom wasn't the one to discipline us. She really had no say in it. (Study Interview #449)

DEALING WITH FAMILY TURMOIL

These women tried to endure the family tension and, given the extended kinship network, some were able to escape this violence by staying with relatives when the situation at home exploded. The majority of them could no longer bear the family violence and actively sought alternatives. Some believed the best strategy for dealing with family violence was to start their own family. Marty, a 34-year-old, describes the process:

> My parents were working. Then in the fifth grade, we moved, and . . . my father got sick, mom had to go on welfare. Things started not working out for the family. My parents was fighting, my father used to give my mother lickings every time and put us down. They were strict. We pretty much rely on each other [the siblings].
>
> I never did get along with my dad. I don't know why. I've always tried, cleaning up, never had to be told what to do, I took care of my sisters and

brothers. Cleaned the house, cook, did all kinds of house chores, but my
father couldn't stand me. . . . I couldn't take it anymore, so I got about to the
seventh grade, that's when I met my husband. I wanted to get married but I
couldn't. So I got pregnant, my first daughter, about a year after that, I quit
school already. I came home, I told my mom I wanted to get married. So she
gave me consent. My father, never. So I forged his name . . . I was 15 years
old. Stayed with my husband and never went back home. Only went back
home once in awhile to give my mom money and see how she doing. (Joe,
1995b, p. 411)

Other women took a path already described in earlier chapters of this book
and ran away from home. Living periodically with friends, relatives, or on the
streets, and sometimes turning to prostitution for survival, these young women
were also fleeing family troubles. Take a look at the following description of
Linda's life:

Linda's parents divorced after her birth, and she has never known her mother.
She and her sister were raised principally by her grandmother. Her father
raped her and her sister, in addition to constantly beating them. The sexual
abuse started when she was nine and continued until she ran at 12 years of
age by "hopping on a bus to Waikiki" and getting lost. She had been in and
out of foster homes and on the streets, but this break was permanent. She
hooked up with a girl in her 20s, "I watched her, she was a prostitute. I asked
her how to do that cause she had a lot of money. She taught me the ropes and
I went for it. I made my money and stayed away from home. I lived out of
hotel rooms." (Study Interview #510)

PATHWAY TO DRUGS

While these women confronted numerous problems—poverty, gendered
expectations and obligations, parental alcohol and illicit drug use, violence,
living on the streets—they found themselves drifting into illicit drug use. Most
of them have regularly used alcohol, tobacco, marijuana, powder cocaine,
crack, and crystal methamphetamine (known in the islands as ice or batu).

Their initiation into drugs usually started with alcohol, marijuana, cocaine,
and then ice. Their peer groups and, most important, family members usually
introduced them to alcohol, tobacco, and marijuana during their early teen
years. Sometimes, the family member was a parent, usually the father or a close

adult relative, such as an uncle. Evie, a 27-year-old woman, remembers the setting when she began smoking marijuana:

> When I was 11. Yeah, my first hit. My first joint. He [father] rolled a joint. Back then, they used to have those little rolling machines and my dad would have ounces of weed in his freezer. So we'd sit there eating ice cream and rolling joints and making bags. Then the boys would come over, hang out. My dad was, he was hanging out, he was involved in underground entertainment so he knew all of the entertainers, all the promoters, artists, drug dealers, he always had hip parties. (Study Interview #401)

The women continued to use alcohol and marijuana with their peers and family members. Eventually, they started using cocaine, initiated typically by other family members, such as siblings, cousins, or other relatives. As their user networks and contacts broadened, women found other sources for their first encounters with ice. Many women tried ice with a small group of their girlfriends. Others were introduced to ice by a relative, typically a cousin or sister-in-law. Several of the younger women indicated that male dealers, who had motives other than a new potential customer, negotiated their first encounter with ice. Finally, approximately a third of them initially tried ice with their partner, and the experience often was associated with sexual enhancement.

Women initially rationalized their ice use in gendered ways (Joe, 1995b). The drug has the effect of an appetite suppressant, which allows them to stay thin and thus provides a source of self-confidence. Moreover, the ice gave them an energy boost that allowed them to transcend and complete the monotonous tasks of "women's work," namely, domestic chores. Several traded "domestic work" with their dealers for drug supplies. Sometimes, their dealer was a relative.

> I started buying from one of my cousins. I used to always burn myself cause I was trying to learn how the hell to do this thing without wasting 'em. My cousin used to see me do that so she taught me . . . I caught on that night! That's when I really felt good! I was up all night long till the next day. . . . I stayed with her for three months. They were big time dealers. They was selling big quantities. I help her clean up the house, a big big house. My auntie's house because I would help her clean and cook, she always used to give me free stash. . . . Right now, the only one supply me is my husband [who does not use]. Then check in one hotel. . . . (p. 411)

With prolonged use, however, they become isolated from others—their children, partners, friends, and families—and at this stage, ice becomes a strategy for coping with their deteriorating situation. This social isolation stems from several sources. Long episodes of limited sleep and food make them irritable. Many respondents spoke of periods of depression and paranoia. The paranoia usually involved feelings that they were being watched and followed by the police and by other users wanting to steal their supplies, and, consequently, they tried to limit their interaction with others. Also, nearly all of them reported weight loss. Some had grown emaciated and exhibited facial sores from tweaking (obsessively focusing on an activity, such as picking at acne) and dehydration. Under these conditions, they limited contact with their family, hoping that relatives would not see their deterioration. Finally, if their partner was using ice, they both became more irritable as a result of lack of sleep and food and money problems. The partner's irritability was often expressed through domestic violence.

Many of these women have become isolated and have a strained relationship with their family, but the extended kinship network allows them to rely on various family members for support in coping with daily life. This includes providing financial support, temporary shelter, and shelter and support for their children. Although this extended kinship system provides these women with consistent support, it has the paradoxical consequence of enabling their drug use, intensifying dependency, and further aggravating family tensions. Take the situation of Stephanie, a 35-year-old:

> While growing up, she recalls that her parents, both alcoholics, began physically beating her at five years of age with "extension cord wires, water hoses, punches, everything." She ran away, and after high school, married and became pregnant. Her husband died shortly after the son's birth in a work-related accident. She has been homeless for seven years, and sometimes stays with friends. Periodically she visits her mother and son, but adds that her ice use has "interfered" with her relationship with her mother. Her mother has been caring for her son since she has "no place for me and my boy." She regularly gives half of her welfare monies to her mother for her son's food and clothing. (Study Interview #475)

Like other women in this study, Stephanie takes refuge in ice as she finds herself with fewer and fewer options. She states, "I can't get no help finding me and my boy a place. So because I'm homeless, that's why I do the drug, I get so depressed cause I don't have no roof over my head for me and my boy." Paradoxically, her family, which caused her to run away, is one of her few remaining resources.

DEMYSTIFYING WOMEN OF COLOR

These women's experiences directly challenge Western notions of the "ideal" Asian American or Pacific Islander family. The "Asian family" is diverse and complex. Moreover, the family cannot be understood in isolation but must be examined in relation to its interaction with gender and class. In this way, we can examine the paradoxical effects of the family as both protector and facilitator of problems such as illicit drug use.

The lives of the women in this study raise a number of important research, prevention, and treatment issues regarding family and class. As we have seen in this sample, the extended family kindles solidarity and cooperation, and when one is financially hard-pressed, it can act as a vital lifeline, a source of stability and solace. At the same time, however, the economic pressures remain. Family conflicts persist and often heighten, particularly for the "traditional breadwinner" of the house. Women who violate cultural norms of femininity further aggravate an already heated home environment. Family conflict may be expressed in many forms: abandonment, neglect, verbal attacks, parental problems with alcohol and drugs, and physical and sexual violence. These women did not respond passively. They sought refuge from the family chaos through the most readily available means. Sometimes, they moved in with other relatives or with friends for as long as possible or lived on the streets.

During this period of chaos, most of these women were at a high risk and began using alcohol and marijuana, sometimes initiated by a male relative. Gradually, their expanded user networks, which often included extended kin (cousins, etc.), initiated them into other illicit drugs. Although they grow increasingly isolated from others because of their extended use of ice, and the family tension from their childhoods remains, the cultural customs of the extended family system paradoxically offer them a source of shelter and support, and a place back home.

GENDER, CULTURE, AND DRUG USE

There is little in these accounts to suggest that Asian Pacific women drug users represent a new, violent woman offender in a degendered environment. Their lives illustrate, if anything, the profound role played by both gender and poverty in their experience with drug use. Their stories illustrate how their culture, particularly those cultural patterns that emphasize extended family lifestyles, tends to function as double-edged swords in their lives. Certainly their family lives produced many problems for them, and yet, at the same time, other elements of this same system provided them with resources and support.

The lives of these women resonate with themes we have seen earlier in this book, particularly the role of violence and victimization in women's deviance and drug use. However, victimization is not the totality of their lives, as these interviews show. Young women responding to severe problems of abuse and neglect seek to escape this violence through any means possible. In the cultures studied here, the extended family offers some respite. For many, however, drugs and, occasionally, the streets are also options. Eventually, these coping strategies produce problems for the women; problems that the family, ironically, steps in to solve where and when it can.

"CRACK PIPE AS PIMP": DRUGS, ETHNICITY, AND GENDER IN AFRICAN AMERICAN COMMUNITIES

One might be tempted to imagine that the patterns described here are unique to women in Hawaii. Looking across the country, however, shows just the reverse. Although the settings and the ethnicity of the girls and women change, the pain and resilience heard in the voices of the women in Hawaii can also be heard in other ethnic communities.

Ethnographic field work conducted in the neighborhoods of Chicago (Ouellet, Wiebel, Jimenez, & Johnson, 1993), Harlem (Bourgois & Dunlap, 1993), and New York (Maher & Curtis, 1992; Maher, Dunlap, Johnson, & Hamid, 1996; Williams, 1992), and interviews with young women in Miami (Inciardi, Lockwood, & Pottieger, 1993), Harlem (Fullilove, Lown, & Fullilove, 1992), Toronto (Erickson, Butters, McGillicuddy, & Hallgren, 2000), and San Francisco (Schwartz et al., 1992) add an important dimension to the discussion of women, culture, and drugs in communities where other ethnic groups (notably African American and Hispanic groups) dominate. These studies also show that the type of drug can change but the damage caused by poverty and racism remains a constant. In most of these studies, it is not ice but crack cocaine that causes the destruction of women's lives, and, once again, women's cultures provide solutions to difficulties that grow out of life on the edge.

These research efforts also break new ground in exploring the specific links between drug use and other forms of women's crime. This body of research also shows that, contrary to timeworn and androcentric images of the male heroin addict, drug use is increasingly becoming polydrug use. Furthermore, women (particularly those who find their way into the criminal justice system) have more problems with drugs than men (Deschenes & Anglin, 1992; Hser, Anglin, & Chou, 1992; Pohl & Boyd, 1992).

Research based on three studies of street youth conducted in Miami during the years 1985 through 1991 (Inciardi et al., 1993) powerfully documents this situation. Two were large-scale interview projects that focused largely on drug use and criminal activity, and one was an ethnographic study of crack houses. The authors' interviews with Miami street girls and women, in particular, reflect a dramatic expansion of the role of crack cocaine in women's lives; about half (49%) of the women reported "current" use of cocaine in the late 1970s, compared to 73% in the early 1980s (Inciardi et al., p. 113).

Bourgois and Dunlap (1993) hasten to caution the reader against demonizing such increases, noting that crack cocaine is simply the "latest medium through which the already desperate are expressing publicly their suffering and hopelessness" (p. 98). Research on the effect of different drugs (alcohol, marijuana, and crystal methamphetamine) among Hawaii's women underscores the need to study the situation of people in poverty, not particular drugs (however horrific).

The study of women drug users in Miami outlines the gendered nature of drug use, noting that although men use drugs for "thrills or pleasure" or in response to peer pressure, women are more likely to drink or use illegal drugs for "self-medication" (Inciardi et al., 1993, p. 25). Women who use drugs, like most women who find their way into the criminal justice system, have histories of extensive physical and sexual victimization, and drug use begins as a way to flee from this pain (see also Fullilove et al., 1992, p. 275). Erickson et al.'s (2000) study of women crack users in Toronto highlights such trauma and crack as a means of escape. To many of these women, quitting crack was not deemed a feasible option because many did not identify their drug use as a problem; crack was what "made their lives interesting or bearable" (p. 777). Stating reasons for what initially led them to trying crack, the women in their study fell into two main explanatory categories: They were either already embedded in drug use when they began using crack or a traumatic event or series of events provided the impetus (p. 773). As one woman stated,

> I was just out of a bad, abusive relationship. I tried to kill myself. . . . I was lonely and depressed. I was watching TV one day, COPS or something, and saw it (crack) being done. I actually drove my car to and asked for a rock. And so I started on my own. (p. 773)

Another woman in their study resonated a similar theme: "It makes me feel good, it makes me happy, it makes me forget all the bad things" (p. 774).

The extent of the violence in girls' and women's lives is also dramatically underscored in interviews with women crack users in Harlem, most of whom are African American. This research focused on the lives of 14 women crack users and documented that "trauma was a common occurrence in their lives," and propelled the women into drug use (Fullilove et al., 1992, p. 277). Like their counterparts in Hawaii, these women's lives were full of the chaos and abuse common in neighborhoods of extreme poverty. They were also distressed and depressed at their inability to "maintain culturally defined gender roles" (specifically, functioning as mothers) because of their drug use. Finally, these women were further victimized by the "male-oriented drug culture," which has developed a bizarre and exploitative form of prostitution around women's addiction to crack cocaine.

Specifically, these neighborhoods have seen the development of a "barter system in which sex—rather than money—can be exchanged for drugs" (Fullilove et al., 1992, p. 276). This system feeds on the particular nature of crack usage, wherein the drug is consumed in periodic binges and "pursuit and use of the drug outweigh other concerns" (p. 276). This pattern of drug use has created the "crack ho"—a prostitute who will trade sex for extremely small amounts of money or drugs, often, but not always, in crack houses generally run by men. Crack addiction, in short, has facilitated the development of a form of prostitution that "may involve participation in bizarre sexual practices for very small amounts of money" and the subsequent "degradation of women within crack culture" (p. 276). The exploitation and subordination of these women goes beyond the mere solicitation of sex. Drug-abusing women surviving in the inner-city crack cultures may experience severe forms of humiliation, as they comply or are forced to participate in sexual acts they believe deviate from normal sexual services (Erickson et al., 2000; Maher et al., 1996). Crack then becomes powerfully useful in eliminating inhibitions and desensitizing these women to the traumatic effects of their sex work and experiences with rape (Young, Boyd, & Hubbell, 2000, p. 796).

An example of the sort of desperation seen in these neighborhoods is supplied by another Harlem researcher (Bourgois), who reported being stopped by a "high school girl" who grabbed him by the arms in a housing project stairwell, "sobbing hysterically" and begging him, "Please! Please! Let me suck you off for two dollars—I'll swallow it. Please! Please! I promise!" (Bourgois & Dunlap, 1993, p. 102). Likewise, Inciardi et al. (1993) observe that women have become the special victims of crack cocaine and provide harrowing stories of sex-for-drug exchanges in crack houses. Here, according to the authors, vulnerable and victimized girls and young women trade sex for extremely small amounts of crack (sometimes as little as $3 worth) and, in the process, expose themselves to the risk of AIDS.

The risk of sexual and physical violence female sex workers face is further highlighted in Pyett and Warr's (1997, 1999) research on female prostitutes in Victoria. The authors find that because of clients' resistance to condom use, physical threats, coercion, and HIV and other STDs remain constant threats in the lives of these women. Drug abuse, inexperience, homelessness, and absence of legal protection increase their vulnerability on the streets. All of the women in their study had experienced violent attacks. Once assaulted, the women often felt that because they were "prostitutes" and worked outside any legally sanctioned modes of sex work, they felt disconnected from social support, community empathy, or processes of justice (1997, p. 545).

These interviews and ethnographies link women's patterns of drug use to women's involvement in prostitution. Because prostitution has always played such a central role in women's crime (Chesney-Lind & Rodriguez, 1983; Miller, 1986), it is important now to turn to the ways in which patterns of prostitution have been affected by the drug scene in certain cities, and how these two trends are, in turn, linked to women's involvement in violent behavior.

PROSTITUTION AND DRUG USE

There is no doubt that the crack scene has affected street prostitution, long a mainstay of women's survival tactics in marginalized communities. In a study of the effect of crack on three Chicago neighborhoods, it was noted that although "street-level prostitution has probably always had rate cutters" (Ouellet et al., 1993, p. 88), the arrival of crack and the construction of the "crack ho" has created a desperate form of prostitution involving instances of extreme degradation that had previously only been seen in extremely impoverished countries such as the Philippines and Thailand (Enloe, 1989; Studervant & Stoltzfus, 1992).

Most women involved in prostitution, even those involved in crack-related prostitution, view their activities as "work" and feel that this work is more ethical and safer than either stealing or drug dealing (Bourgois & Dunlap, 1993, p. 104). According to this perspective, because the crack epidemic has virtually destroyed the economic viability of street prostitution in some neighborhoods, some women might even prefer to go to these houses to get their drugs directly rather than risk the dangers of the streets and violence from johns who are even less well-known to them than the men in the crack houses.

Not all women addicted to crack choose this path, particularly when times get hard. For this reason, crack has also changed the nature of street prostitution in neighborhoods where it is present. Maher and Curtis's (1992) work on women involved in street-level sex markets in New York, for example, notes that the introduction of crack cocaine "increased the number of women working the strolls and had a significant impact on the kind of work they did, the renumeration they received and the interactions that occurred in and around street-level sex markets" (p. 21). Maher and Curtis found that women involved in prostitution were also involved in other forms of property crime, such as shoplifting, stealing from their families, and occasionally robbing johns as a way to survive on the streets.

Increased competition among women involved in prostitution and the deflation in the value of their work has created a more hostile environment among New York streetwalkers, and an increased willingness to rip off johns (see also Ouellet et al., 1993). To understand this, it is important to convey the enormity of the violence that women in sex work are routinely exposed to at the hands of johns. Maher and Curtis (1992) provide many such examples. One woman they talked to told them,

I got shot twice since I bin here . . . was a car pulled up, two guys in it, they was like "C'mon gon on a date." I wouldn't go with that so they came back arund and shot me . . . in the leg and up here. (p. 23)

Another told them,

I got punched in the mouth not too long ago they [two dates] ripped me off—they wanted their money back after we finished—threw me off the van naked—then hit me with a blackjack 'cause I jumped back on to the van because I wanted to get my clothes. It was freezing outside so I jumped back onto the van to try and get my clothes and he smacked me with a blackjack on my eye. (p. 244)

In fact, some "robbery" becomes much more understandable when seen up close. Take Candy's story:

I robbed a guy up here not too long ago—5 o'clock Sunday morning . . . a real cheek gonna tell me $5 for a blojob and that pisses me off—arguing wit them. I don't argue no more—jus get in the car sucker, he open his pants and do like this and I do like this, put my hand, money first. He give me the money I say, "See ya, hate to be ya, next time motherfucker it cost you $5 to get me to come to the window." (Maher & Curtis, 1992, p. 246. Used by permission)

Although women are clearly using crack, Bourgois and Dunlap (1993, p. 123) found little evidence that women were involved in selling the drug. Inciardi et al. (1993) found slightly more evidence, noting that women (including many prostitutes) are drawn to selling crack due to its availability and low cost (unlike heroin). All agree that prostitution is still a mainstay for street women, as is petty property crime. In contrast, female violent crime and major property crime is far less frequent—even among these women—and clearly related to "heavy" crack use (Inciardi et al., p. 120).

VICTIMIZATION, PROSTITUTION, AND WOMEN'S CRIME

Attempting to chart girls' and women's journeys into crime invariably brings one back to the role of prostitution in women's offenses. For example, when

Miller (1986) sought to undertake a study of women felons in Milwaukee, she initially excluded women arrested only for prostitution (because it was only a misdemeanor in Wisconsin). Even with this rule in place, she found that "it soon became clear that almost everyone I interviewed had at least one arrest for prostitution as well" (p. 25).

Likewise, a study of women in prison in Hawaii (Chesney-Lind & Rodriguez, 1983) found that of women interviewed (two thirds of whom were sentenced felons at the time), the majority (88%) of the women had been involved with prostitution.

Understanding runaway girls' decisions to engage in survival sex while on the streets is relatively easy. But why do adult women, even those who are not seriously addicted to drugs, continue to be involved in sex work? The answer is both sad and understandable. Interviews with women in Milwaukee and Hawaii, and the New York work by Maher and Curtis, link this decision to women's survival needs. Prostitution, it is often observed, does not provide women with "easy" money, but it is "fast" money. However, in the words of one former prostitute, "fast money doesn't last" (K. Hill, personal communication, April 11, 1996).

Women's comments indicated that the most common reason they started working as prostitutes was financial. In Hawaii, women were recruited directly out of what might be called bar-related female professions into prostitution. Half the women were working off and on in entertainment or bar-related jobs while they were involved in prostitution—indicating that these occupations, in themselves, serve as "adjuncts rather than alternatives to female criminal activity" (Chesney-Lind & Rodriguez, 1983, p. 55). James (1976), in her work on prostitution in Seattle, cites two studies that connect the employment in "occupations . . . that adhere most closely to the traditional female service role, often emphasizing physical appearance as well as service" to the entrance of women into prostitution (p. 188). Employers often require female employees in service jobs to "flirt with customers and 'be sexy,'" which frequently results in the women finding that the men they must serve at their jobs already consider them to be "no better than a prostitute" (James, p. 188).

Drug use and addiction often becomes a problem for women involved in prostitution, even for those who were not initially involved in prostitution to support their drug habits. Another Hawaii study (Chesney-Lind & Rodriguez, 1983) found that the need to buy drugs was a factor in women's involvement in theft, burglary, and robbery, for which they were also more likely to be incarcerated than for prostitution. In the interviews with these women, it was obvious that, for most, their entry into prostitution predated their heavy drug use. However, drug dependency, perhaps developed as part of life "in the fast lane," quickly made their exit from the profession unlikely and quite probably encouraged them to seek even more money through burglary and theft (p. 57).

Daly (1994) found a more varied pattern of women's entry into serious crime in her content analysis of probation officers' reports on 40 women appearing in New Haven's felony court. Four categories of women offenders emerged from these accounts: "street women" (who were sexually abused, ran away as girls, and got involved in prostitution), "harmed and harming women" (who were abused or neglected as children, are probably drug addicted, and are likely to get violent when under the influence), "battered women" (who are in a relationship with a violent man and act in response to this battery), and "drug-connected women" (who are using and selling drugs in connection with boyfriends or family members; Daly, p. 47). Of these four categories, however, the first two accounted for the bulk of women's cases, again underscoring the importance of victimization and drug use in women's more serious offending.

Another Hawaii study focused exclusively on women and on one violent crime—robbery. In this research, Kauffman (1993) found that women accounted for just 9% of those arrested for this offense in 1991 in Honolulu. This confirms that, although women's robbery arrests explain the maintenance of the violent crime rate for females over time (Chapman, 1980), robbery remains an overwhelmingly male offense.

This research also found that the criminal careers of women arrested for robbery in Hawaii were significantly different from those of male robbers. Specifically, women were less likely than men to have been arrested for serious crimes of violence. Instead, most of the women's other arrests, unlike those of the male robbers, were for the traditional female offenses, and many were related directly or indirectly to prostitution. These include forgery or credit card fraud and prostitution and its related offenses of disorderly conduct and pedestrian violations.[7]

Kauffman (1993) found that over a third (39%) of the female robbers but only 19% of the men had previous arrests for "disorderly conduct," and 35% of the women but only 4% of the men had prior arrests for "pedestrian violations" (often a buffer charge for prostitution; p. 23). Women in this study were also about as likely as men to have a prior history of drug arrests. A careful review of these arrest histories led Kauffman to conclude that, of women arrested for the violent crime of robbery in Hawaii in 1991, at least 43% had arrest histories that indicated they were probably prostitutes (p. 28).

Prostitution is often called the "oldest profession," which is a sad but telling comment about the stability of girls' and women's work—legal and illegal. Many have argued that it is a "victimless" crime because the customers and the prostitute are knowing and willing participants in the act. This chapter has provided evidence, however, that the phenomenon can set the stage for violence against women and that the prostitute herself sees little of the true profits of her labor. Moreover, in part because she sees so little of the money due her, she is tempted to use the role of prostitute to rip off violent and callous johns (K. Hill, personal communication, April 11, 1996).

Finally, discussing prostitution and women's violence responsibly is impossible without recognizing the gendered nature of the lives and work of the prostitutes themselves. As we have seen, women involved in drug use, prostitution, and other forms of women's crime have experienced victimization directly related to their gender in the form of incest, sexual abuse, and rape. Ironically, they have conformed to society's gender role expectations in career aspirations and in relationships and, as a result, the victimization related to their gender continues in their adult relationships with both pimps and customers. In those crack cultures in which women users appear to be freed from oppressive pimping structures as they work competitively and independently, the gender relations within these new crack culture contexts still amplify gender inequalities, continue women's dependence on men, and create new extremes of sexual exploitation and victimization (Bourgois & Dunlap, 1993; Maher et al., 1996).

CONCLUSION

The work on women's entry into criminal behavior, taken altogether, illuminates the ways in which the injuries of girlhood produce problems that young women often solve on the streets of poor neighborhoods. That the straight and illicit economies the women find on those streets is gendered is made very clear in these studies. In addition, it is evident that violence is a part of life in these communities, that women have always been exposed to large amounts of violence, and that women are capable of responding in ways that can be categorized as "violent." Generally, it has served the interests of the powerful to ignore or minimize women's ability to engage in violence (White & Kowalski, 1994). After all, as noted earlier, given the amount of violence women suffer at male hands, the remarkable story is that women are not more violent.

Campbell (1993) notes the androcentric perspective on women's violence is that "violent women must be either trying to be men or just crazy" (p. 144). What this research illustrates is the importance of placing a woman's violence within the totality of her life. Then, it becomes clear that "the problem of female aggression is located within interpersonal and institutionalized patterns of a patriarchal society" (White & Kowalski, 1994, p. 502).

Finally, prostitution, however renamed and reshaped, remains the major gateway to women's entry into other forms of illicit activities because girls' and women's "capital" is still chiefly their sexuality. What is not clear is that drugs play a more central role in the prostitution of the 1990s than they did in earlier decades (Bourgois & Dunlap, 1993), when women's drug addiction and crime went largely ignored and only the prostitution caused concern. What has changed, though, is the public response to the drug addiction and the violence that have probably always surrounded street life. This change, to be considered in the next chapter, has had a huge and unanticipated effect on the nation's criminal justice system, because it has caused a dramatic surge in women's imprisonment.

NOTES

1. The sex/gender system has three components, according to Renzetti and Curran (1995): (a) the social construction of gender categories based on biological sex, (b) a sexual division of labor in which specific tasks are allocated based on sex, and (c) the social regulation of sexuality in which particular forms of sexual expression are either positively or negatively sanctioned (pp. 2–3).

2. An earlier version of this section was drawn from Joe (1996a).

3. A comprehensive discussion of existing substance abuse research on Asian Pacific Americans can be found in Zane and Sasao (1992). The methodological problems involved in conducting research on social problems such as illicit drug use and delinquency among Asian Pacific populations can be found in Joe (1993, 1996b).

4. Part of the data stem from a larger National Institute on Drug Abuse (NIDA) study of adult moderate to heavy methamphetamine users in three locales noted for high usage and problems with this drug—Honolulu, San Francisco, and San Diego

(Morgan et al., 1994). Consistent with emergency admissions and treatment reports, the Honolulu site was the only one that included a significant number of Asian Pacific American users. Three fourths (or 111) of the 150 active users interviewed in Hawaii were Asian Pacific American. Approximately a third ($n = 37$) of the 111 Asians in Hawaii were women. Women accounted for one third of the total number of users in all three sites. A second study using a similar ethnographic design is being conducted by the second author with 30 female users of ice or crack cocaine.

5. A more extensive discussion of ohana is found in Joe (1995b). See also Pukui, Haertig, and Lee (1972).

6. Samoan gender relations revolve around Polynesian traditions of male dominance, separation, and obligation. Although Hawaiian customs were similar to the Polynesian model of separation, this was severely altered with the death of Kamehamehakunuiakea in 1819 and the subsequent arrival of the missionaries. Although the Hawaiian system retains some features of male dominance, it is the women who have "learned the ways of the malihini (strangers). Women adjusted to and became clever at cultural and economic transactions with the new world" (Nunes & Whitney, 1994, p. 60). At the same time, however, Hawaiians, who are the most marginalized group in the state, have acclimated to poverty through normalizing early motherhood, high dropout rates, and welfare dependency for girls (see Joe & Chesney-Lind, 1995). In modern Filipino families, girls and women have been socialized according to colonial cultural and religious, usually Catholic, norms that emphasize the secondary status of women, girls' responsibility to their families, and the control of female sexual experimentation (Aquino, 1994).

7. Comparing those women with previous arrests for prostitution with those without revealed that prostitutes had higher rates of offending and a higher average number of arrests for robbery, assault, theft/larceny-theft, weapons violations, and drug-related, alcohol-related, and prostitution-related offenses. Of those not arrested for prostitution, 23% had been arrested for offenses typically related to prostitution. Those not arrested for prostitution had a slightly higher rate of arrests for burglary and a larger number of arrests for forgery/credit card fraud.

SEVEN

SENTENCING WOMEN TO PRISON

Equality Without Justice

———◦•◦———

The number of women imprisoned in the United States has increased sixfold in the past two decades, bringing the number of women behind bars to over 93,000.[1] From 1985 to 1995 alone, women's overall correctional status more than doubled in the United States, with the number of women in prison or on parole increasing threefold (see Table 7.1). By the end of the century, women accounted for 16% of the total corrections population, and over a third of them served time in the nation's three largest jurisdictions: Texas, the Federal system, and California. (Bureau of Justice Statistics, 1988, 1999, 2002a; see also Chesney-Lind, 2002, pp. 80–81).

Increases in the number of women in prison surpassed those of men over this period, as well. In the 1990s, the number of women in prison rose by 110%, compared to men's increase of 77%. Between 1995 and 2001, the average annual increase in the state and federal correctional population for women was 5.2% and for men was 3.7%. The incarceration rate rose 23% for women during this time, whereas men's incarceration rate grew by only 13% (Bureau of Justice Statistics, 2002a, p. 6). A similar pattern was found in the number of women in jail, where a 32% increase was seen between 1995 and 2000; for men, the increase was 21% (Bureau of Justice Statistics, 2001).

Table 7.1   One Woman in _____ Had This Correctional Status in the
United States

Year	Any Correctional Status	Probation	Jail	Prison	Parole
1985	227	267	4762	4167	4762
1990	161	202	2632	2326	2273
1995	124	159	1961	1587	1493

SOURCE: Bureau of Justice Statistics. (1999). *Women Offenders* (p. 6). Washington, DC: U.S. Department of Justice.

The soaring numbers of women under lock and key are not simply products of the increasing reliance in the United States on imprisonment, although that has played a role in the pattern. Women's proportional "share" of the total prison population has also increased. In 1984, for example, women accounted for 7% of those in jail (Maquire & Pastore, 1994, p. 591), 4.4% of those in state prisons (Bureau of Justice Statistics, 1985, p. 4), and 5.8% of those in federal penitentiaries. Fifteen years later, women accounted for 11% of those in jail (Bureau of Justice Statistics, 2001, 2002a), 6.7% of those in state prisons, and 7% of those in federal institutions. Additionally, between 1990 and 2000, the number of women on probation and parole grew by 8% (Bureau of Justice Statistics, 2002b).

Nationally, the rate of women's imprisonment is also at an all-time high. In 1925, women's rate of incarceration was 6 per 100,000. By 2001, the rate climbed to 58 per 100,000, with Hispanic and African American women experiencing even higher rates of incarceration (see Table 7.2). Taken together, these figures signal a major policy change in society's response to women's crime, one that has occurred with virtually no public discussion.

So, as the number of people imprisoned in the United States continues to climb, our nation has achieved the dubious honor of having the highest incarceration rate in the world, with Russia closely following as second (Mauer, 1999). Along the way, America's love affair with prisons has claimed some hidden victims—economically marginalized women of color and their children.

TRENDS IN WOMEN'S CRIME: A REPRISE

Is the dramatic increase in women's imprisonment a response to a women's crime problem spiraling out of control? As seen in previous chapters, a look at

Table 7.2   Number of Sentenced Female Prisoners Under State or Federal
Jurisdiction per 100,000 Residents by Race, Hispanic Origin, and
Age, 2001

Age	Total	White	Black	Hispanic
Total	58	36	199	61
18-19	31	25	83	23
20-24	91	61	225	105
25-29	164	94	483	150
30-34	211	130	682	176
35-39	165	102	561	147
40-44	88	51	320	88
45-54	42	27	136	61
55 and older	6	5	18	7

SOURCE: Bureau of Justice Statistics. (2002a). *Prisoners in 2001* (p. 12). Washington, DC:
U.S. Department of Justice.

the pattern of women's arrests provides little evidence of a dramatic change in
the composition of women's crime. One crude measure will serve to make this
point again. The total number of arrests of adult women (which might be seen
as a measure of women's criminal activity) increased by 15% between 1992
and 2001 (FBI, 2002, p. 239). However, the number of women incarcerated
during the last 5 years alone increased by 75%, despite decreases in most
violent crime offenses (Bureau of Justice Statistics, 2002a).

WOMEN, VIOLENT CRIMES,
AND THE WAR ON DRUGS

Another indication that the increase in women's imprisonment is not explained
by a shift in the character of women's crime comes from information about the
offenses for which women are being imprisoned (see Table 7.3). Whereas 55%
of the increase in men's incarceration rate is due to violent offenses, only 33%
of the growth in women's incarceration is due to violent crimes. The proportion of women in state prisons for violent offenses declined from 48.9% in 1979
to 32% in 2001 (Bureau of Justice Statistics, 1988, 2002a). In states that have
seen large increases in women's imprisonment, such as California, the decline
is even sharper. In 1992, only 16% of the women admitted to the California
prison system were being incarcerated for violent crimes, compared to 37.2%
in 1982 (Bloom, Chesney-Lind, & Owen, 1994).

Table 7.3   Partitioning by Gender and Offense, the Growth of the Sentenced
                  Prison Population Under State Jurisdiction, 1990–2000

Offense	Total Increase/Percent of Total	Male Prisoners Increase/Percent of Total	Female Prisoners Increase/Percent of Total
Total	516,800/100	477,300/100	39,700/100
Violent	273,200/53	260,300/55	12,900/33
Property	63,500/12	56,000/12	7,500/19
Drug	101,400/20	88,500/19	12,900/33
Public order	78,800/15	72,500/15	6,300/16

SOURCE: Bureau of Justice Statistics. (2002a). *Prisoners in 2001* (p. 13). Washington, DC:
U.S. Department of Justice.

Other recent figures suggest that without any fanfare, the "war on drugs"
has become a war on women and has contributed to the explosion in women's
prison populations. One out of three women in U.S. prisons in 2001 was
doing time for drug offenses (up from 1 in 10 in 1979), whereas one out of
five men were imprisoned for drug convictions (Bureau of Justice Statistics,
2002a, p. 13; Snell & Morton, 1994, p. 3). Although the intent of "get tough"
policies was to rid society of drug dealers and so-called kingpins, over a third
(35.9%) of the women serving time for drug offenses in the nation's prisons
are serving time solely for "possession" (Bureau of Justice Statistics, 1988,
p. 3).[2]

The war on drugs, coupled with the development of new technologies for
determining drug use (e.g., urinalysis), plays another, less obvious role in
increasing women's imprisonment. Many women parolees are being returned to
prison for technical parole violations because they fail to pass random drug tests.
Of the 6,000 women incarcerated in California in 1993, approximately a third
(32%) were imprisoned for parole violations. In Oregon, during a 1-year period
(October 1992–September 1993), only 16% of female admissions to Oregon
institutions were incarcerated for new convictions; the rest were probation and
parole violators. This pattern was not nearly so clear in male imprisonment; 48%
of the admissions to male prisons were for new offenses (Anderson, 1994).
Finally, in Hawaii, recent data underscore this point further: Of individuals
released during 1998 and tracked for 2 years on parole, nearly half (43%) were
returned to prison. When examining the reasons for parole revocation, a gender
difference emerges: 73% of the women were returned to prison for technical

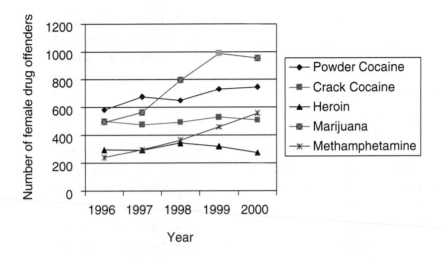

Figure 7.1 Female Drug Offenders Sentenced in Federal Court by Year and Drug Type, 1996–2000

SOURCE: United States Sentencing Commission, fiscal years 1997–2001 datafiles.

Note: Percentage change, 1996–2000:

Powder cocaine:	+28%
Crack Cocaine:	+.02%
Heroin:	−.07%
Marijuana:	+93%
Methamphetamine:	+133%

violations (as opposed to new crimes); this was true for a smaller, yet significant, percentage (64%) of male parolees (Chesney-Lind, 2002, p. 90).

Nowhere has the drug war taken a larger toll than on women sentenced in federal courts. In the federal system, the passage of harsh mandatory minimums for federal crimes, coupled with new sentencing guidelines intended to "reduce race, class and other unwarranted disparities in sentencing males" (Raeder, 1993) have operated to the distinct disadvantage of women.[3] They have also dramatically increased the number of women sentenced to federal institutions. In 1989, 44.5% of the women incarcerated in federal institutions were being held for drug offenses. Only 2 years later, this was up to 68%. Drugs that have come under amplified surveillance by the federal government within the last decade, such as methamphetamine and marijuana, have also greatly impacted women. From 1995 through 2000, the number of women convicted of federal methamphetamine offenses increased by 133% (from 239

offenders in 1996 to 558 in 2000), and female federal marijuana offenders
were up by 93% (from 495 in 1996 to 953 in 2000; see Figure 7.1). The
number of women convicted of powder cocaine offenses increased by 28%,
and the number of female crack cocaine offenders remained at a steady high
of about 500 offenders.

Additionally, drugs such as methamphetamine and crack cocaine come
attached with mandatory prison sentences for relatively small amounts of drug
trafficking. The consequence is that more and more women no longer receive
probation for low-level offenses but, rather, receive prison. Twenty years ago,
nearly two thirds of the women convicted of federal felonies were granted
probation, but in 1991 only 28% of women were given straight probation
(Raeder, 1993, pp. 31–32). The mean time to be served by women drug
offenders increased from 27 months in July 1984 to a startling 67 months
in June 1990 (p. 34). Taken together, these data explain why the number of
women in federal institutions has skyrocketed since the late 1980s. In 1988,
before full implementation of sentencing guidelines, women made up 6.5% of
those in federal institutions; by 1992, this figure had jumped to 8% and
remained at this percentage into the 21st century. From 1988 to 1992, the
number of women in federal institutions climbed by 97.4% (Bureau of Justice
Statistics, 1989, p. 4; 1993, p. 4) and has grown by an average 7% annually
since then (Bureau of Justice Statistics, 2002a, p. 7).

What about property offenses? Over 25% of the women in state prisons
were doing time for these offenses in 2001. California, again, merits a closer
look: Over a third (34.1%) of women in California state prisons in 1993 were
incarcerated for property offenses of which "petty theft with a prior offense"
is the most common offense. This generally includes shoplifting and other
minor theft. One woman in 10 in California prisons is doing time for petty
theft. Taken together, this means that one woman in four is incarcerated in
California for either simple drug possession or petty theft with a prior (Bloom,
Chesney-Lind, & Owen, 1994, p. 3).

GETTING TOUGH ON WOMEN'S CRIME

Data on the offenses for which women are in prison and an examination of
trends in women's arrests suggest that factors other than a shift in the nature of
women's crime are involved in the dramatic increase in women's imprisonment.

Simply put, the criminal justice system now seems more willing to incarcerate women.

What has happened in the last decade? Although explanations are necessarily speculative, some reasonable suggestions can be advanced. First, it appears that mandatory sentencing for specific kinds of offenses—especially drug offenses—at both state and federal levels has affected women's incarceration. Legislators at the state and national level, perhaps responding to a huge increase in media coverage of crime but not necessarily the nation's actual crime rate, are escalating penalties for all offenses, particularly those associated with drugs ("Crime Down," 1994; Mauer & Huling, 1995).

Beyond this, sentencing "reform," especially the development of sentencing guidelines and mandatory minimums resulting from "Three Strikes and You're Out" legislation, also has been a problem for women. In California, this has resulted in increasing the number of prison sentences for women (Blumstein, Cohen, Martin, & Tonry, 1983). Sentencing reform has created some problems because the reforms address issues that have developed in the handling of male offenders and are now being applied to female offenders.[4] Daly's (1991) review of this problem notes, for example, that federal sentencing guidelines ordinarily do not permit a defendant's employment or family ties/familial responsibilities to be used as a factor in sentencing. She notes that these guidelines probably were intended to reduce class and race disparities in sentencing, but their effect on women's sentencing was not considered. Bush-Baskette's (1999) analysis of the war on drugs resonates a similar theme: "Sentencing guidelines that disallow the use of drug addiction and family responsibilities as mitigating circumstances subject Black females to prison and long sentences under criminal justice supervision, as they do White females" (p. 222).

Finally, the criminal justice system has simply become tougher at every level of decision making. Langan (1991) notes that the chance of receiving a prison sentence following arrest has risen for all types of offenses, not simply those typically targeted by mandatory sentencing programs (p. 1569). This is specifically relevant to women because mandatory sentencing laws (with the exception of those regarding prostitution and drug offenses) typically have targeted predominantly male offenses, such as sexual assault, murder, and weapons offenses. Thus, Langan's research confirms that the whole system is now "tougher" on all offenses, including those that women traditionally have committed.

A careful review of the evidence on the current surge in women's incarceration suggests that this explosion may have little to do with a major change in women's behavior. This surge stands in stark contrast to the earlier growth in women's imprisonment, particularly to the other great growth of women's incarceration at the turn of the 20th century.

Perhaps the best way to place the current wave of women's imprisonment in perspective is to recall earlier approaches to women's incarceration. Historically, women prisoners have been few in number and were apparently an afterthought in a system devoted to the imprisonment of men. In fact, early women's facilities were often an outgrowth of men's prisons. In those early days, women inmates were seen as "more depraved" than their male counterparts because they were viewed as acting in contradiction to their whole "moral organization" (Rafter, 1990, p. 13).

The first large-scale, organized imprisonment of women occurred in the United States when many women's reformatories were established between 1870 and 1900. Women's imprisonment was justified not because the women posed a public safety risk, but because women were thought to need moral revision and protection. Important to note, however, is that the reformatory movement that resulted in the incarceration of large numbers of white working-class girls and women for largely noncriminal or deportment offenses did not extend to women of color. Instead, as Rafter (1990) has carefully documented, African American women, particularly in the southern states, continued to be incarcerated in prisons where they were treated much like the male inmates. They frequently ended up on chain gangs and were not shielded from beatings if they did not keep up with the work (pp. 150–151). This racist legacy, the exclusion of black women from the "chivalry" accorded white women, should be kept in mind when the current explosion of women's prison populations is considered.

The current trend in adult women's imprisonment seems to revisit the earliest approach to female offenders: Women are once again an afterthought in a correctional process that is punitive rather than corrective. Women are also, however, no longer being accorded the benefits, however dubious, of the chivalry that had characterized the reformatory movement. Rather, they are increasingly likely to be incarcerated, not because society has decided to crack down on women's crime specifically, but because women are being swept up in a societal move to "get tough on crime" that is driven by images of violent criminals (almost always male and often members of minority groups) "getting away with murder."

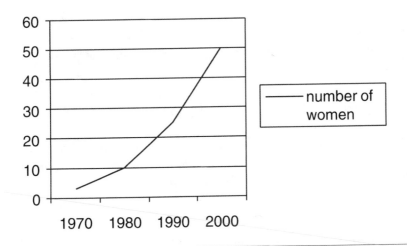

Figure 7.2   Number of Women Under the Death Sentence in the United States, 1970–2000

SOURCE: Bureau of Justice Statistics (1999, 2002a).

A look at capital punishment demonstrates this point further. Although historically women have received the death penalty far less frequently than men, the advent of "get tough" approaches to crime ushered in a dramatic increase in the number of death sentences imposed on women (Morgan, 2000, p 280; see Figure 7.2). During the 1990s, the total number of women sentenced to death exceeded the combined total of the previous two decades, despite declines in murder arrests for women. From 1998 to 2001, seven women were executed in the United States, the most women executed in a 3-year period during the past 100 years (Streib, 2002).

This public mood, coupled with a legal system that now espouses "equality" for women with a vengeance when it comes to the punishment of crime, has resulted in a much greater use of imprisonment in response to women's crime. There also seems to be a return to the imagery of women's depravity for those women whose crimes (and race) put them outside of the ranks of "true women." As evidence, consider the new hostility signaled by bringing child abuse charges against women who use drugs before the birth of their children (Chavkin, 1990; Noble, 1988).

The fact that many of the women incarcerated in U.S. prisons are women of color who are doing time for drug offenses further distances them from images of womanhood that require protection from prison life. For this reason, when policymakers are confronted with the unanticipated consequences of the new "get tough" mood, their response is all too frequently to assail the character of the women they are jailing rather than to question the practice itself.

BUILDING MORE WOMEN'S PRISONS

As a result of the surge in women's imprisonment, our country has gone on a building binge with regard to women's prisons. Prison historian Nicole Hahn Rafter (1990) observes that between 1930 and 1950, roughly two or three prisons were built or created for women each decade. In the 1960s, the pace of prison construction picked up slightly, with seven units opening, largely in southern and western states. During the 1970s, 17 prisons opened, including units in states such as Rhode Island and Vermont, which once relied on transferring women prisoners out of state. In the 1980s, 34 women's units or prisons were established; this figure is 10 times larger than the figures for earlier decades (Rafter, pp. 181–182).

To put this dramatic shift in another important historical context, consider the fact that only 30 years ago, the majority of states did not operate separate women's prisons. In 1973, only 28 states (including Puerto Rico and the District of Columbia) had separate institutions for women. Other states handled the problem differently; women were either housed in a portion of a male facility or, like Hawaii, Rhode Island, or Vermont, were imprisoned in other states (Singer, 1973). Looking backward, this pattern was very significant. The official response to women's crime during the 1970s was heavily influenced by the relative absence of women's prisons, despite the fact that some women were, during these years, committing serious crimes.

What has happened in the past few decades, then, signals a major and dramatic change in the way the country is responding to women's offenses. Without much fanfare and with little public discussion, the model of men's incarceration has been increasingly applied to women. Some of this punitive response to women's crime can be described as "equality with a vengeance"—the dark side of the equity or parity model of justice that emphasizes the need to treat women offenders as though they were "equal" to male offenders. As one correctional officer said at a national meeting, "an inmate is an inmate is an inmate."

But who are these "inmates," and does it make sense to treat women in prison as though they were men? The next section examines what is known about the backgrounds of women currently doing time in state and federal prisons across the country.

PROFILE OF WOMEN IN U.S. PRISONS

Childhoods of Women in Prison

The most recent research on the characteristics of women doing time in state prisons across the country underscores the salience of themes identified early in this book—particularly the role of sexual and physical violence in the lives of women who come into the criminal justice system. This research also argues forcefully for a national discussion of the situation of women in our jails and prisons.

Snell and Morton (1994) surveyed a random sample of women and men ($N = 13,986$) in prisons around the country during 1991 for the Bureau of Justice Statistics. For the first time, a government study asked questions about women's and men's experiences of sexual and physical violence as children.

They found, when they asked these questions, that women in prisons have experienced far higher rates of physical and sexual abuse than men. Forty-three percent of the women surveyed "reported they had been abused at least once" before their current admission to prison; the comparable figure for men was 12.2% (Snell & Morton, 1994, p. 5). A look at women in jail and prison in 1998 shows even higher estimates: 48% of women in jail and 57% in state prisons report prior histories of sexual or physical abuse (see Table 7.4).

For about a third of all women in prison (31.7%), the abuse started when they were girls and continued as they became adults. A key gender difference emerges here. A number of young men who are in prison (10.7%) also report abuse as boys, but it did not continue into adulthood. One in four women reported that their abuse started as adults, compared to only 3% of male offenders. Fully 33.5% of the women surveyed reported physical abuse, and a slightly higher number (33.9%) had been sexually abused either as girls or young women, compared to relatively small percentages of men (10% of boys and 5.3% of adult men in prison).

This survey also asked women about their relationships with those who abused them. Predictably, both women and men reported that parents and relatives contributed to the abuse they suffered as children, but female prisoners were far more likely than their male counterparts to say that domestic violence

Table 7.4   Characteristics of Adult Women on Probation, in Jail, and in Prison

	Probation	Jails	State Prisons
Race/Ethnicity			
White	62%	36%	33%
Black	27	44	48
Hispanic	10	15	15
Other	1	5	4
Age			
24 and younger	20	21	12
25–34	39	46	43
35–44	30	27	34
45–54	10	5	9
55 and older	1	1	2
Median Age	**32 years**	**31 years**	**33 years**
Marital Status			
Married	26	15	17
Widowed	2	4	6
Separated	10	13	10
Divorced	20	20	20
Never married	42	48	47
Education			
8th grade or less	5	12	7
Some high school	35	33	37
High school graduate/GED	39	39	39
Some college or more	21	16	17
Report ever physically or sexually abused	**41**	**48**	**57**

SOURCE: Bureau of Justice Statistics. (1999). *Women Offenders* (pp. 7–8). Washington, DC: Department of Justice.

was a theme in their adult abuse; fully half of the women said they had been abused by a spouse or ex-spouse, compared to only 3% of male inmates.

The survey found ethnic differences in the role played by the juvenile and criminal justice system in the lives of women in prison. Overall, white women were slightly more likely to report having been in the foster care system or other institutions (21.1%) than African American or Hispanic women (14.1% and 14.4%, respectively). African American women and Hispanic women, by contrast, were far more likely than white women to report a family member (usually a brother) in prison (Snell & Morton, 1994).

Contrary to some stereotypes about drug use, more white and Hispanic women than African American women reported parental involvement with alcohol and drug abuse when they were girls. Over 4 out of 10 white women, and about a third of the Hispanic women, reported parental drug abuse, compared to only a quarter of the African American women. This underscores the need to focus on the specific interaction among culture, gender, and class in women's pathways to prison.

Current Offenses

A look at the offenses for which women are incarcerated quickly puts to rest the notion of hyperviolent, nontraditional women criminals. "Nearly half of all women in prison are currently serving a sentence for a nonviolent offense and have been convicted in the past of only nonviolent offenses" (Snell & Morton, 1994, p. 1). In fact, the number of women in prison for violent offenses, as a proportion of all female offenders, has fallen steadily over the past decades, whereas the number of women in prison has soared. In 1979, about half of the women in state prisons were incarcerated for violent crimes (Bureau of Justice Statistics, 1988). By 1986, the number had fallen to 40.7%, and in 2001 it was at 32.2% (Bureau of Justice Statistics, 2002a; Snell & Morton, p. 3). One out of three women in U.S. prisons is there for a violent crime, compared to about one out of every two male prisoners.

Snell and Morton (1994) also probed the gendered nature of the women's violence that resulted in their imprisonment. They noted that women prisoners were far more likely to kill an intimate or relative (50%, compared to 16.3%), whereas men were more likely to kill strangers (50.5%, compared to 35.1%). More recent data show similar findings. In 1998, over 93% of female homicide offenders killed an intimate, family member, or acquaintance. For men, only 76% killed someone they knew (see Table 7.5). Given the information already discussed in this book regarding the nature of women's violence and its relationship to their own histories and experiences of abuse, women's violent acts take on quite a different significance than men's violence.

Drugs and their role in women's violence are also apparent in these data; generally speaking, women doing time for crimes of violence were less likely to report a link between drugs and violence than women serving time for property or drug offenses. For example, only 11% of the women convicted of

Table 7.5   Relationship of Offender to Victim for Murder Offenses, 1998

Victim	Female	Male
Spouse	28.3%	6.8%
Ex-spouse	1.5	0.5
Child/stepchild	10.4	2.2
Other family member	6.7	6.9
Boyfriend/girlfriend	14.0	3.9
Acquaintance	31.9	54.6
Stranger	7.2	25.1
Number, 1976–1997	59,996	395,446

SOURCE: Bureau of Justice Statistics. (1999). *Women Offenders* (p. 4). Washington, DC: Department of Justice.

violent crimes used drugs at the time of their crime, compared to 25% of those serving time for property offenses, and 32% of those serving time for drug offenses (Bureau of Justice Statistics, 1999, p. 9). In Snell and Morton's (1994) study, the one exception to this generalization was found for women incarcerated for robbery. Not only did these women report that they were under the influence of the drug at the time of the robbery, but they were virtually the only women serving time for a violent offense who reported that they committed the offense "to get money for drugs" (p. 8). Women serving time for homicide were also slightly more likely to report greater use of drugs the month before the offense for which they were imprisoned and to report being under the influence of drugs at the time of the offense, but they rarely said that getting money to buy drugs was a motive for the crime.

Property Crimes

Many women in state prisons are serving time for larceny theft. Indeed, of the women serving time for property offenses (25.1% of all women in prison), about a third (30.2%) are doing time for larceny theft. This compares to only 18.2% of men who are doing time for property crimes. Fraud is another important commitment offense for women, accounting for 40% of women's, but only 11.3% of men's, most serious property offenses. Men serving time for property offenses are more likely to be serving time for burglary (49.2%; Bureau of Justice Statistics, 2002a).

Drug Use Among Women in Prison

Given the past history of the women in prison, it should come as no surprise that drug use, possession, and, increasingly, drug trafficking are themes in women's imprisonment. In 1979, only 10.5% of women in state prisons were serving time for drug offenses; by 1986, the proportion had increased to 12%, but in 2001, nearly a third (32.3%) of all women in state prisons were doing time for drug offenses (Bureau of Justice Statistics, 1988, p. 3; 2002a, p. 13; Snell & Morton, 1994, p. 3). Currently, 57% of the women serving time for drug offenses are now serving time for drug trafficking. Although this offense sounds very serious, it must be placed in context. As we shall see later in this chapter, in a world where big drug deals are controlled almost exclusively by men (Green, 1996), women, many of whom are from desperately poor countries or from our own impoverished communities, are being cast or coerced into the role of serving as drug mules or couriers, only to be swept up in the escalating penalties that have characterized the past decade's war on drugs (Mauer & Huling, 1995).

National data on women in prison confirm that women prisoners have more problems with drugs than their male counterparts. Snell and Morton (1994) found that, contrary to the stereotype of the male drug addict committing crimes, "women in prison in 1991 used more drugs and used those drugs more frequently than men" (p. 7). For example, more female prisoners used drugs daily before imprisonment than male prisoners (41.5%, compared to 35.7%), and women were more likely than men to be under the influence of drugs when they committed the offense for which they were imprisoned (36.3%, compared to 30.6%). Finally, about a quarter of women in prison but only a fifth of men committed the offense for which they were imprisoned to buy drugs.

Ominously, about a quarter of all women in prison had some form of drug treatment prior to imprisonment, and of those using drugs, 41.8% had treatment the month before their offense. These figures suggest that most interventions are not sufficient to help these women with their drug problems.

Women prisoners are also taking health risks by using drugs. Snell and Morton (1994) found that women prisoners were more likely than men to use needles to inject drugs (34%, compared to 24.3%) and to have shared needles with friends (18%, compared to 11.5%). Again, contrary to many stereotypes, these rates were highest among white and Hispanic women, compared to African American women. For example, 41.6% of white women and 45.9% of Hispanic women had ever used a needle, compared to only 24% of African American women.

Perhaps as a result of these patterns, at the end of 2000 more women than men in prison were infected with HIV; in that year, 3.6% of all women in state prisons had the virus that causes AIDS, compared to 2.2% of male inmates. In New York, the state with the most HIV-positive female prisoners (600), the percentage of women inmates testing positive (18.2% of the female prison population) far outreached the percentage of male inmates testing positive for the virus (8%) (Maruschak, 2002). Although the number of inmates with HIV is decreasing overall, the rate is faster for men. Between 1999 and 2000, 7% fewer men in prison were known to be HIV positive; for women, the decrease was 2% (Maruschak, 2001, 2002).

Mothers Behind Bars

Of the 869,600 women under some form of criminal justice surveillance, over 70% have children under the age of 18; this accounts for over 1.3 million children (Bureau of Justice Statistics, 1999; see Table 7.6). Many of these women will never see their children if this and other national studies (see Bloom & Steinhart, 1993) are accurate. Snell and Morton (1994) found that 52.2% of the women with children under 18 had never been visited by their children. Most of the women who were able to be visited by their children saw them "less than once a month" or "once a month." More women were able to send mail to or phone their children, but even here, one in five never sent or received mail from their children, and one in four never talked on the phone with their children. This is despite the fact that many of these women, prior to their incarceration, were taking care of their children (unlike their male counterparts).

According to Snell and Morton (1994), just under three quarters of the women with children had lived with them before going to prison, compared to only slightly over half (52.9%) of the male prisoners. Moreover, because women's work is never done, it is more often the imprisoned woman's mother (the child's grandmother) who takes care of her children, whereas male inmates are more likely (89.7%) to be able to count on the children's mother to care of the child (Snell & Morton, p. 6).

Table 7.6   Children of Women Under Correctional Supervision, 1998

	Women Offenders	Women Offenders With Children Under Age 18	Number of Minor Children
Probation	721,400	516,200	1,067,200
Jail	63,800	44,700	105,300
State prisons	75,200	49,200	117,100
Federal prisons	9,200	5,400	11,200
Total	869,600	615,500	1,300,800

SOURCE: Bureau of Justice Statistics. (1999). *Women Offenders* (p. 7). Washington, DC: Department of Justice.

These patterns are particularly pronounced among African American and Hispanic women. White female inmates more often report access to husbands as primary caretakers of their children, whereas African American women do not identify this as an option (Enos, 2001, p. 55). Although black women and Hispanic women are more likely to share caretaking responsibilities with other family members and are less likely to rely upon foster care services, the ability of the family to effectively respond, both financially and emotionally, to the incarceration of a female family member with children is dependent upon social and economic status (Enos). This becomes highly problematic for women of color, because poverty and race are intertwined and families often have few resources to extend.

Race and Women's Imprisonment

Snell and Morton's (1994) study, like those done before, clearly documents the number of women of color behind bars. The numbers indicate that more than half the women in the nation's prisons are African American (46%) and Hispanic (14.2%). Hidden in these data is the fact that the surge in women's imprisonment has disproportionately hit women of color in the United States. Further analysis of these survey data and other national data (Mauer & Huling, 1995) has thoroughly documented the way in which the surge in women's imprisonment has been driven almost completely by a dramatic increase in the imprisonment of women of color. Although white women comprise 62% of women on probation, it is African American women who are most represented in jails and prisons (see Table 7.4).

Between 1986 and 1991, all women saw an increase in what Mauer and Huling (1995) call the "control rate" (the proportion of women under some form of correctional supervision—probation, jail, prison, or parole), but this rate jumped most dramatically for African American women. Although much of the nation's attention has been correctly focused on the horrific overcontrol of African American males (whose control rate now approaches one out of every three young males between the ages of 20 and 29; Mauer & Huling), their sisters are also seeing increases in contact with the criminal justice system.

The "control rate" for African American women was 2.7% of all young women in 1989; by 1994, the rate had jumped 78% to 4.8% (or 1 out of 20 young African American women; Mauer & Huling, 1995, p. 5). The distance between the white and African American rates also widened, so that well over three times as many young black women have contact with the criminal justice system than do their white counterparts. Hispanic women have also seen their "control rate" increase by 18%, and their "control rate" is about double the rate for white women (2.2%).

Mauer and Huling (1995) present compelling evidence to support their contention that much of this increase can be laid at the door of the war on drugs, which many now believe has become a war on women, particularly on women of color. They also present a striking analysis of how the crackdown on drug use and trafficking has affected black and Hispanic women. Specifically, although the number of women in state prisons for drug sales has increased by 433% between 1986 and 1991, this increase is far steeper for Hispanic women (328%) and for African American women (828%) than for white women (241%; Mauer & Huling, p. 20).

Huling (1995), in a subsequent paper, directly links these increases in women's incarceration to the fact that the war on drugs has been particularly harsh on those using and selling crack cocaine. This has a significant effect on African American women because "there are indications that women are more likely to use crack and are more likely to be involved in crack distribution sales relative to other drugs" (p. 8). Thus, she contends that, without much public fanfare, the war on drugs, and particularly the harsh penalties for the sale of crack cocaine (relative to powder cocaine and other drugs), has had a dramatic effect on the incarceration patterns of African American women. With the major focus of the drug war on low-level street users of crack cocaine, black women, constructed by the media as "crack whores" and drug-addicted mothers, became "responsible" for crack's devastation in inner-city neighborhoods (see also Bush-Baskette, 1999, for similar argument). Consequently, black women entered the criminal justice system at exacerbated rates.

Recall the research by English (1993) on women's and men's self-reported drug selling, wherein she found that female prisoners were much more likely than their male counterparts to report numerous small drug sales. This could mean that the patterns of women's drug selling, rather than the seriousness of their sales, expose them to more risk of arrest and incarceration.

The other hidden victims of the war on drugs are the women, many from foreign countries, who are serving time in U.S. prisons for being "drug couriers." Huling (1996) notes that the lack of repatriation treaties between most "drug-demand countries" and "drug-supply countries" has meant that many drug couriers end up serving long prison terms in the country of their arrest. Initially, women from foreign countries entering the United States at airports, such as John F. Kennedy in New York, were tried in federal court. As the federal prisons began to experience sharp increases in women's imprisonment, federal officials shifted the cases to state courts (Huling, 1996; see also English, 1993).

Reviewing the cases of women who were arrested at JFK airport during 1990 and 1991 for drug smuggling ($N = 59$), Huling (1996) found the following: First, almost all (96%) had no history of involvement with the criminal justice system. Most (95%) had not been convicted at trial but had instead plead guilty to a reduced charge. To avoid the New York laws that would have sentenced them to life terms, they plead guilty to a reduced charge that requires a "mandatory minimum" of 3 years to life in prison. Almost all of the women arrested were Hispanic (Huling, p. 53). Prosecutors, when asked about these patterns, argued that they had "no choice" but to pursue indictments for anyone found in possession of 4 ounces or more of an illegal drug.

Interviewing some of these women, Huling (1996) was able to document that many carried the drugs because of threats to their families, because they were trapped in abusive relationships with men involved in the drug trade, or because they had been duped or fooled. Despite this reality, Huling shows that New York politicians (including elected prosecutors) used the number of convictions of drug smugglers to document their "get tough on crime" stances, and despite a public outcry generated in part by Huling's work and the work of Sister Marion of Riker's Island, efforts to reform New York's harsh mandatory sentences failed.

Different Versus Equal?

Given situations like those experienced by women charged with being drug smugglers, it should come as no surprise that the continuing debate over whether equality under the law is a good thing for women has special immediacy for those looking at the situation of women in the criminal justice system. To recap this debate (see Chesney-Lind & Pollock-Byrne, 1995, for a full discussion), some feminist legal scholars argue that the only way to eliminate the discriminatory treatment and oppression that women have experienced in the past is to push for continued equalization under the law; that is, to champion equal rights amendments and to oppose any legislation that treats men and women differently. They argue that although this may hurt in the short run, in the long run it is the only way that women will ever be treated as equal playing partners in economic and social spheres. For example, MacKinnon (1987) writes, "For women to affirm difference, when difference means dominance, as it does with gender, means to affirm the qualities and characteristics of powerlessness" (pp. 38–39). Even those who do not view the experience of women as one of oppression conclude that women will be victimized by laws created from "concern and affection" that are designed to protect them (Kirp, Yudof, & Franks, 1986).

The opposing argument is that women are not the same as men and because it is a male standard that equality is measured against, they will always lose. Therefore, one must consider differential needs (a sort of separate but equal argument). This would mean that women and men might receive differential treatment as long as it did not put women in a more negative position than the absence of such legislation. Conversely, the equalization proponents feel that, given legal and social realities, differential treatment for women will always be unequal treatment and by accepting different definitions and treatments, women run the risk of perpetuating the stereotype of women as "different from" and "less than" men.

One might reasonably ask how this legal debate, which has to date largely focused on the rights of women as workers, bears on women as prisoners. In fact, as the next section will demonstrate, the experience of women prisoners starkly illuminates some of the shortcomings of the conventional extremes of the different versus equal debate, because at different points in our nation's history, those who have imprisoned women have used each perspective to deal with the women they confined. This review of the history and current issues surrounding women's imprisonment will also highlight severe problems with the gender-blind approach to jailing women.

Prisons and Parity

Initially, the differential needs approach was the dominant correctional policy. Almost from the outset, the correctional response to women offenders was to embrace the Victorian notion of "separate spheres" and to construct and manage women's facilities based around what were seen as immutable differences between men and women (Rafter, 1990). Women were housed in separate facilities and programs for women prisoners represented their perceived role in society. Thus, they were taught to be good mothers and housekeepers; vocational education was slighted in favor of domestic training. Women were hired to supervise female offenders in the belief that only they could provide for the special needs of female offenders and serve as role models for them. To some degree, this legacy still permeates women's prisons today. Typically, these prisons have sex-typed vocational programming and architectural differences (such as smaller living units and decentralized kitchens) in recognition of gender roles.

In sentencing, too, one could observe that the system treated women and men differently. Women were much less likely than men to be imprisoned unless the female offender did not fit the stereotypical female role, for example, if she was a "bad mother" who abused or abandoned her children, or if she did not have a family to care for (Chesney-Lind, 1987; Eaton, 1986). This resulted in one of the most dramatic disproportional ratios in criminal justice—women composed roughly only 4% of the total prison population for years. Of course, part of this was because most women committed far fewer serious crimes than men, but at least some part of the difference was due to sentencing practices (see Blumstein et al., 1983).

Certainly, the differential treatment of women in sentencing and prison programming is a thing of the past. Partially as a result of prisoner rights' litigation based on the parity model (see Pollock-Byrne, 1990), women offenders are being swept into a system that seems bent on treating women "equally." Currently, the emphasis on women's prison construction and the architecture of women's prisons suggest that women get the worst of both worlds, correctionally. A couple of well-publicized scandals can serve to highlight the severe problems with a "gender-blind" approach to women's imprisonment.

In Alabama, the state reinstated male "chain gangs" with much fanfare in 1995, after they had been dropped in 1932 because of accounts of brutality and abuse. The current practice involves men shackled in groups of five working along public highways, although some groups are assigned the job of breaking "large rocks into little ones" ("Chain Gang Death," 1996, p. 8A). The country's current "get tough on crime" mood provided Alabama officials with the opening to reinstate these workgroups and even to involve some groups in grueling "busy work."

Alabama corrections officials were threatened with a lawsuit brought by male inmates suggesting that the practice of excluding women from the chain gangs was unconstitutional. The response from the Alabama Corrections Commissioner was to include women in the chain gangs (Franklin, 1996, p. 3A). Ultimately, the governor forced the corrections chief to resign; the governor's spokesperson said simply, "It was just a philosophical difference. In his [the governor's] opinion, there is a difference in men and women (specifically) physically" (Hulen, 1996, p. A1).

Although the issue of chain gangs for women is moot in Alabama, it has surfaced in other states. Proclaiming himself "an equal opportunity incarcerator," an Arizona sheriff has started one for women "now locked up with three or four others in dank, cramped disciplinary cells" (Kim, 1996, p. 1A). To escape these conditions, the women can "volunteer" for the 15-woman chain gang. Defending his controversial move, the sheriff commented, "If women can fight for their country, and bless them for that, if they can walk a beat, if they can protect the people and arrest violators of the law, then they should have no problem with picking up trash in 120-degrees" (p. 1A).

Other routine institutional practices, such as strip searches (sometimes involving body-cavity searches), have also produced problems. In New York prisons, in response to complaints by male inmates that strip searches were often accompanied by beatings, video monitors (usually mounted on the wall) were installed in areas where searches occurred. When the women's prison (Albion Correctional Center) began to tape women's strip searches, though, fixed cameras were replaced by hand-held cameras (Craig, 1995, p. 1A).

Fifteen women prisoners incarcerated at Albion filed complaints based on their experiences with strip searches. Specifically, they said that doors to the search area were occasionally kept open, that male guards were sometimes seen outside the doors watching the searches, and that, unlike the men's videos, which surveyed the whole room where the searches occurred, according to the women's lawyer, "these videotapes were solely focused on the woman. That amplified the pornographic effect of it" (Craig, 1996, p. 1A). Said one woman who was searched while men were "right outside a door and could see the whole incident, 'I knew they was watching . . . I was so humiliated . . . I felt like I was on display. I felt like a piece of meat'" (p. A6). Advocates for the women stressed the traumatic effect of such searches, given the histories of sexual abuse and assault that many women bring with them to prison.

Moreover, the women inmates suspected that prison officials were viewing the tapes and eventually filed complaints to stop routine videotaping of women prisoners. In addition to receiving over $60,000 in damages, the women were able to change the policy of routine videotaping of women's searches. As a result of their complaints, "a female inmate would be filmed only if officers believed she would resist the search" (Craig, 1996, p. A6). Presently, very few searches of women inmates are being videotaped in New York, but the possibility of abuse is present in almost all prisons.

Even without videotaping and other possible abuses, strip searches have quite different meanings for women and men. For example, a key point made by the Albion women was that, given the high levels of previous sexual abuse among women inmates, such searches had the possibility of being extremely traumatic. In fact, similar concerns have also surfaced in a Task Force Report to the Massachusetts Department of Health about the use of "restraint and seclusion" among psychiatric patients who have histories of sexual abuse (Carmen et al., 1996).

Finally, the most pervasive complaint that has accompanied women's imprisonment is the sexual abuse and harassment of women inmates by male guards. Owen's (1998) study of "women in the mix" includes the degrading experiences women encounter with male guards. Although the women in her study offered limited discussion about forced or consensual sexual relationships with staff, one woman illustrated the potential for harassment in day-to-day activities:

> If you are short, the officers, you can be seen from the (officer's) bubble. It
> is degrading. Sometimes you get a shower peeker. I told the other girls to
> block the shower. Then the officer got an attitude. You could tell. (p. 166)

As old as women's imprisonment (Beddoe, 1979), the sexual victimization of women in U.S. prisons is the subject of increasing news coverage and, more recently, international scrutiny. Scandals have erupted in California, Georgia, Hawaii, Ohio, Louisiana, Michigan, Tennessee, New York, and New Mexico (respectively, Stein, 1996; Meyer, 1992; Watson, 1992; Curriden, 1993; Sewenely, 1993; Craig, 1996; and Lopez, 1993), and the assumption has grown that prisons, here in the United States and elsewhere, are rife with this problem. So extensive is the concern that the issue has attracted the attention of Human Rights Watch (1993).

Details of these scandals yield the predictable charges and counter-charges, but the storyline remains essentially unchanged; women in prisons, guarded by large numbers of men, are vulnerable. As one advocate for women in prison has noted, "We put [women] into an environment where they're controlled by men and men are willing to put their hands on them whenever they want to" (Craig, 1996, p. A1). The story that prompted this observation dealt with one of a series of sexual assaults reported by a young woman in a New York prison:

> Correctional officer Selbourne Reid, 27, came into the cell of a 21-year-old inmate at the maximum security prison in Westchester County. The inmate, at first asleep, was startled to find him in her cell. . . . On this night, Reid forced the woman to perform oral sex on him, according to the Westchester County district attorney's office. After he left, she spit the semen into a small bottle in her room. She told prison authorities about the attack, and gave them the semen for DNA analysis. (Craig, 1996, pp. 1A, 6A)

Similar accounts appear with distressing regularity, and even more disturbing is the fact that so few of the cases, unlike the one reported here, actually go to trial or result in the perpetrators being found guilty. Institutional subcultures in women's prisons that encourage correctional workers to "cover" for each other, coupled with inadequate protection accorded women who file complaints, make it unlikely that many women inmates will show the courage of the young woman in New York. Indeed, according to a memo filed by an attorney in the Civil Rights Division of the U.S. Department of Justice, the Division found "a pattern of sexual abuse by both male and female guards" in Michigan women's prisons (Patrick, 1995).

A judge reviewing the situation of women in Washington D.C. jails noted that "the evidence revealed a level of sexual harassment which is so malicious that it violates contemporary standards of decency" (Stein, 1996, p. 24). If this is true, why do so few of these cases make it to court? Sadly, some of this involves the histories of women in prison, many of whom have engaged in prostitution, which allow the defendants to use the misogynist defense that it is impossible to rape a prostitute. Beyond this, the public stereotype of women in prison as "bad girls" means that any victim must first battle this perception before her case can be fairly heard. Finally, what little progress has been made is severely threatened by recent legislation that has drastically curtailed the ability of prisoners and advocates to sue over prison conditions (p. 24)— changes again likely to have been motivated by public perceptions of prisoners as violent men.

That women in prison are the recipients of "equity with a vengeance" does not necessarily mean that the abuses that used to exist in prisons that assumed gender difference have retreated completely. In fact, it appears that today's women in prison still receive some of the worst of the old separate spheres abuses, particularly in the area of social control. For example, McClellan (1994) examined disciplinary practices at men's and women's prisons in Texas. Using Texas Department of Corrections records, McClellan constructed two samples of inmates (271 men and 245 women) and followed them for a 1-year period (1989). She documented that although most men in her sample (63.5%) had no citation or only one citation for a rule violation, only 17.1% of the women in her sample had such clear records. Women prisoners were much more likely to receive numerous citations and received them for different sorts of offenses than men. Most commonly, women were cited for "violating posted rules," whereas men were cited most frequently for "refusing to work" (McClellan, 1994, p. 77). Finally, women were more likely than men to receive the most severe sanctions, including solitary confinement (p. 82).

McClellan's (1994) review of the details of women's infractions subsumed under the category "violation of posted rules" included such offenses as "excessive artwork ('too many family photographs on display'), failing to eat all the food on their plates, and for talking while waiting in the pill line" (p. 85). Possession of contraband could include such things as an extra bra or pillowcase, peppermint sticks, or a properly borrowed comb or hat. Finally, "trafficking" and "trading" included instances of sharing shampoo in a shower and lighting another inmate's cigarette (p. 85).

The author concluded by observing that there exists "two distinct institutional forms of surveillance and control operating at the male and female facilities . . . this policy not only imposes extreme constraints on adult women but also costs the people of the State of Texas a great deal of money"

(McClellan, 1994, p. 87). Research like this provides clear evidence that women in prison are overpoliced and overcontrolled in institutional settings— a finding earlier researchers have noted, as well (see Burkhart, 1973; Mann, 1984). Whether this is an extension of historic interests in women's sexual behavior or whether, more prosaically, it is a function of the fact that, if men were controlled to the extent women were, they would probably riot, is unclear.

What is clear from all these accounts is that women in modern prisons may be subjected to the "worst of both worlds." If McClellan's (1994) findings can be extended to other states, women in modern prisons continue to be overpoliced and overcontrolled (a feature of the separate-spheres legacy of women's imprisonment). At the same time, they are also the recipients of a form of "equality" that results in abuses that are probably unparalleled in male institutions (e.g., sexual exploitation by guards and degrading strip searches). Beyond this, correctional leaders are, in some cases, implementing grossly inappropriate and clearly male-modeled interventions, such as chain gangs and even boot camps, to deal with women's offenses (Elis, MacKenzie, & Simpson, 1992).

The enormous and rapid increase in women's imprisonment has clearly overwhelmed correctional officials who must scramble to come up with space, let alone programs, for the thousands of women coming through the doors (see Morash & Bynum, 1996). Yet prior to these huge population increases, things were not necessarily good for women inmates. Women inmates have never had the same range of programs as male offenders (this was often justified by their low numbers; see Pollock-Byrne, 1990). Because the current imprisonment boom has affected men's and women's facilities, even with larger numbers in women's prisons, women's special needs are unlikely to receive serious attention any time soon.

Some efforts have been made nationally, especially in improving the connection of children with their incarcerated mothers. For example, in California, some nonviolent female drug offenders are sentenced to Family Foundations, a community-based, residential drug-treatment program, where they live with their children who are six and younger. The Women's Prison Association in New York assists women offenders in addressing the critical issues involved in women's pathways toward crime and in their successful return to the community after prison; these issues include substance abuse problems, victimization experiences, family disruption, housing needs, and vocational and employment issues (Conly, 1998). Lastly, The Children's Center in Belford Hills, NY, allows women offenders to reside with their children until the children are 1 year of age. The women learn "to be good mothers," and the focus is on the women's mental health needs (National Institute of Justice, 1998, p. 8).

Despite programs such as these, there still exists a paucity of alternative and innovative approaches available to address women offenders' issues. A National Institute of Justice (1998) study demonstrates this point: In the study, state and prison-level administrators were asked to identify innovative programs for women in prison in their jurisdictions. Only three states reported high levels of innovative programming for women; 34 states identified none or limited availability (National Institute of Justice, p. 6). On a more global level, given the differences between male and female prisoners, it seems extremely unlikely that women's experience of imprisonment will ever mirror men's experience—no matter how often the legal system insists on a gender-neutral stance. Nor, if the lessons are learned from these scandals, should women be treated as though they are men.

The abuses mentioned earlier force us to ask whether a gender-blind approach to imprisonment is fair or just. Is it the case that female prisoners are "disappearing" politically, in a country haunted by images of male drug king-pins and violent predators, because their convictions bolster those who are cynically manipulating the system and the public's fears to win an election? Finally, as the nation becomes increasingly aware of the surge in women's imprisonment from news accounts (LeBlanc, 1996), we need to question whether tax dollars spent on women's imprisonment could be better spent on programs for women in the community.

REDUCING WOMEN'S IMPRISONMENT THROUGH EFFECTIVE COMMUNITY-BASED STRATEGIES AND PROGRAMS

The expansion of the female prison population has been fueled primarily by increased rates of incarceration for drug offenses, not by commitments for crimes of violence. The majority of women in America's prisons are sentenced for nonviolent crimes that are all too often a direct product of the economic marginalization of the women who find their way through the prison doors.

As we have seen, changes in criminal justice policies and procedures over the past decade have contributed to the dramatic growth in the female prison population. Mandatory prison terms and sentencing guidelines are gender-blind and, in the crusade to get tough on crime, criminal justice policymakers have gotten tough on women, pushing them into jails and prisons in unprecedented numbers.

Most of these female offenders are poor, undereducated, unskilled, victims of past physical or sexual abuse, and single mothers of at least two children. They enter the criminal justice system with a host of unique medical, psychological, and financial problems.

The data summarized in this chapter suggest that women may be better served in the community because of the treatable antecedents and less serious nature of their crimes. A growing number of states are beginning to explore nonincarcerative strategies for women offenders, such as the ones aforementioned. Commissions and task forces charged with examining the effect of criminal justice policies on women are recommending sentencing alternatives and the expansion of community-based programs that address the diverse needs of women who come into conflict with the law.

In California, the Senate Concurrent Resolution (SCR) 33 Commission on Female Inmate and Parolee Issues examined the needs of women offenders. The Commission's upcoming report is based on three central concepts: (a) Female inmates differ significantly from males in terms of their needs, and these gender-specific needs should be considered in planning for successful reintegration into the community; (b) women are less violent in the community and in prison, and this fact provides opportunities to develop nonprison-based programs and intermediate sanctions without compromising public safety; and (c) communities need to share the responsibility of assisting in this reintegration by providing supervision, care, and treatment of women offenders (Bloom et al., 1994).

Overcrowding and overuse of women's prisons can be avoided by planning creatively for reduced reliance on imprisonment for women. Many advocate a moratorium on the construction of women's prisons and a serious commitment to the decarceration of women. They believe that every dollar spent locking up women could be better spent on services that would prevent women from resorting to crime. As one prisoner at the Central California Women's Facility commented,

> You can talk to them about community programs. I had asked my P.O. for help—but his supervisor turned him down. I told him that I was getting into a drinking problem, asked if he could place me in a place for alcoholics but

he couldn't get permission. I was violated with a DUI—gave me eight months. I think people with psychological problems and with drug problems need to be in community programs. (Bloom et al., 1994, p. 8)

There are a range of effective residential and nonresidential community-based programs serving women offenders throughout the nation. Austin, Bloom, and Donahue (1992) reviewed limited program evaluation data and found the following common characteristics that appeared to influence successful program outcomes: continuum of care design, clearly stated program expectations, rules and sanctions, consistent supervision, diverse and representative staffing, coordination of community resources, and access to ongoing social and emotional support. They also suggested that promising approaches are multidimensional and deal specifically with women's issues.

DETENTION VERSUS PREVENTION

The United States now imprisons more people than at any time in its history and has the world's highest incarceration rate (Mauer, 1999). On any given day, over a million people are locked up, and an unprecedented number of prison cells are being planned. As a result, the fastest-growing sector of state and local economies, nationally, is correctional employment, which increased 108% during the past decade, whereas total employment increased by just 13.5% (Center for the Study of the States, 1993, p. 2). Women in conflict with the law have become the hidden victims of the nation's imprisonment binge. Women's share of the nation's prison population, measured in either absolute or relative terms, has never been higher. Women were 4% of the nation's imprisoned population shortly after the turn of the 20th century. By 1970, the figure had dropped to 3%. By 2001, however, more than 6.7% of those incarcerated in state prisons in the country were women.

Is this increase in women's imprisonment being fueled by a similarly dramatic increase in serious crimes committed by women? The simple answer is no. As has been shown, the proportion of women in prison for violent crimes has dropped steadily, and the numbers of women incarcerated for petty drug and property offenses have soared. Large increases in women's imprisonment are due to changes in law-enforcement practices, judicial decision making, and legislative mandatory sentencing guidelines, rather than a shift in the nature of the crimes women commit.

As a nation, we face a choice. We can continue to spend our shrinking tax dollars on the pointless and costly incarceration of women guilty of petty drug and property crimes, or we can seek other solutions to the problems of drug-dependent women. Because so many of the women in prison in California are driven to drug use because of poverty and abuse, the real question before us is: detention or prevention?

As this and previous chapters have indicated, we know what to do about crime, particularly crime committed by women. Any review of the back-grounds of women in prison immediately suggests better ways to address their needs. Whether it be more funding for drug treatment programs, more shelters for the victims of domestic violence, or more job training programs, the solutions to their problems are obvious. The question remains: Do we as a society have the courage to admit that the war on drugs (and indirectly on women) has been lost and at a great price (see Baum, 1996)? The hidden victims of that war have seen their petty offenses criminalized and their personal lives severely disrupted. Is this our only choice?

This book has suggested another choice. By focusing on strategies that directly address the problems of women on the economic and political margins rather than expensive and counterproductive penal policies, the pointless waste of the nation's scarce tax dollars could be stopped. To do this, there must be changes in public policy, so that the response to women's offenses addresses human needs rather than the short-sighted objectives of lawmakers who often cannot see beyond the next sound bite or election. The greed of what might be called the "correctional industrial complex" must also be addressed. This term refers to those who benefit from prison construction (such as architectural and construction companies, unions representing prison guards, etc.), who might well seek to replace the mindless spending of the cold war with the equally mindless but profitable incarceration of the nation's poor and dispossessed.

Now that we have entered the new millennium, there are actually a few indications that some states are beginning to reexamine their incarceration practices. So, although the rate of women's imprisonment does stand at an historic high, the first year of the new century saw the first year in which the female rate of increase in imprisonment fell behind that of the male rate of increase (1.2% compared to 1.3%; Beck & Harrison, 2001, p. 5).

Several states long associated with the women's imprisonment boom—notably California and New York—actually saw decreases in the number of women in their prisons. In California, the decrease that accelerated in 2001 was clearly tied to the passage of Proposition 36. This initiative, passed in 2000, diverted most people convicted of nonviolent drug possession to programs instead of prison. In the short time since its inception, it has caused the number of women sent to California prisons to drop by 10% (Martin, 2002, p. 1) The drop actually encouraged two Democratic lawmakers to propose closing one or two of California women's prisons in an attempt to address the state's budget deficit (p. 1).

California's experience provides a valuable lesson to the rest of the nation. Given the characteristics of the women in prison, it is clear that the decarceration of almost all of the women in United States prisons would not jeopardize public safety. Furthermore, the money saved could be reinvested in programs designed to meet women's needs, which would enrich not only their lives but the lives of many other women who are at risk for criminal involvement. Finally, by moving dollars from women's prisons to women's services, we will not only help women—we also help their children. In the process, we are also breaking the cycle of poverty, desperation, crime, and imprisonment, rather than perpetuating it.

NOTES

1. According to Bureau of Justice Statistics (2001, 2002a), during 2001, 85,031 women were held in state and federal correctional facilities, and the "average daily population" of U.S. jails included 161,200 adult women.

2. In 1979, 26% of women doing time in state prisons for drug offenses were incarcerated solely for possession (Bureau of Justice Statistics, 1988, p. 3).

3. Raeder (1993) notes, for example, that judges are constrained by these federal guidelines from considering family responsibilities, particularly pregnancy and motherhood, which in the past may have kept women out of prison. Yet the effect of these "neutral" guidelines is to eliminate from consideration the unique situation of mothers, especially single mothers, unless their situation can be established to be "extraordinary." Nearly 90% of male inmates report that their wives are taking care of their children; by contrast, only 22% of mothers in prison could count on the fathers of their children to care for the children during the mother's imprisonment (p. 69). This means that many women in prison, the majority of whom are mothers, face the potential, if not actual, loss of their children. This is not a penalty that men in prison experience.

4. Blumstein et al. (1983) note that California's Uniform Determinate Sentencing Law "used the averaging approach, one consequence of which was to markedly increase the sentences of women—especially for violent offenses" (p. 114).

◄ EIGHT ►

CONCLUSION

————•◦•————

These are the things I think about at night: 1) will i every be a
normal person again 2) will I ever stop doing drugs 3) do i really
Forgive my Father 4) will i ever love a man 5) will I always be gay
6) Do i betrayed my mother and my sister 7) will i ever stop push-
ing people away From me who care for me 8) will i ever stop pros-
tituting 9) do I really love myself or can i 10) will i ever stop
comming to jail 11) will i ever stop being a criminal & robbing
12) will i ever be me again

> —*Letter from Trina, a prostitute, in*
> *Riker's Island prison for drug possession and*
> *loitering (LeBlanc, 1995)*

T ucked outside of Fresno, California, is the nation's largest prison for
women (LeBlanc, 1996, p. 35). Opened in 1990, the Central California
Women's Facility (CCWF) sits among flat fields planted with nut trees and
growing vegetables near a small, rural town called Chowchilla. CCWF is also
arguably the world's largest women's prison, and yet many in Fresno do not
know it is there (Owen, 1998).

In May 1995, CCWF, with a design capacity of 2,000, housed 3,596
women (79.8% over capacity; Bloom, Chesney-Lind, & Owen, 1994). As a
result of persistent problems with overcrowding at all its women's facilities,

California built a new women's prison of even greater capacity (2,200)—across the street from CCWF. Called Valley State Prison, this near-mirror-image of CCWF is the nation's second-largest prison for women (LeBlanc, 1996, p. 35).

This book has attempted to describe the circumstances that would bring a young woman to Central California Women's Facility and to the other new women's prisons that now dot the country. That a woman, especially a "good" woman, might find her way into one of these institutions is, for most Americans, unthinkable. One hopes that, after reading this book, this comfortable assumption has been challenged.

This book has argued that girls' troubles create and set the stage for girls' and, ultimately, women's crime. Girls' pathways into crime, even into violence, are affected by the gendered nature of their environments and particularly their experiences as marginalized girls in communities racked by poverty. The increase in girls' participation in gangs has, as we have seen, roots in the violence the girls in these communities suffer. Sadly, though, the gang that promised safety and a sense of belonging provides no such haven. Instead, the gang often becomes a new site for girls' exploitation while facilitating their further involvement in violence and crime.

Not all the girls who are arrested, however, are low-income girls of color. As we have seen, sexual abuse of children, unlike physical abuse, knows no class or racial boundaries, and almost all the girls in the juvenile justice system share this terrible and all-too-often secret scar. The links between childhood victimization (both physical and sexual) and efforts undertaken by girls to escape abuse by running away are clear. However, we as a society continue to criminalize these girls' survival strategies, despite nearly three decades of efforts to deinstitutionalize status offenders, including runaways.

Judicial and parental resistance to deinstitutionalization initiatives invites comparison between the situations of the runaway girls today and the situation of runaway wives at the beginning of the last century, when adult women suffered "civil death" after marriage. As a result of having an almost complete absence of legal rights, a woman's property became her husband's. She had to ask her husband for permission to travel to visit her family or friends, and divorce was virtually impossible to secure. If a woman ran away from a brutal husband, not even her own family could legally harbor her (Sinclair, 1956). Abuses of this sort of power fueled the long march to guarantee adult women civil rights during the first wave of feminism. Unfortunately, the civil rights of young people are still severely circumscribed, making the arrest of girls, some of whom are seeking to escape the same problems, not only possible but normal.

After the arrest, the juvenile justice system has increasingly evolved into a two-track system—one for white girls and another for girls of color. White girls are swelling the ranks in private "facilities" and hospitals or placed in social welfare settings where their rights are suspended while they are "helped" and "cured."

As sinister as these patterns are for white girls, the situation for girls of color is much worse. Here, African American girls, Hispanic girls, and Native American girls find themselves in public detention centers and training schools for offenses far less serious than those committed by boys. Additionally, they spend more time in these facilities than do their white counterparts. This racialized pattern of juvenile justice runs parallel to the still prevalent sexism that has haunted the court since its outset, and casts even more doubt as to whether American girls can find justice in such a system.

The role of race and violence in adult women's crime is similarly transparent if one looks at the background of "unruly" women offenders. As we have seen, the women who are filling U.S. prisons share with their counterparts in the juvenile justice system terrible histories of sexual and physical abuse. In their lives, we see that the violence that characterized their girlhoods has followed them into adulthood. In a terrible irony, revictimization, often in the forms of sexual assault and domestic violence, is a common theme in the lives of the adult woman who are arrested, jailed, and imprisoned.

There is no mystery, then, why adult women use drugs. Unlike their patterns of use as girls, where drug use might have been recreational, their involvement with drugs as adults is a mix of self-medication and economic survival (in the form of petty drug sales). The drugs used to push out the pain have become, themselves, huge problems for adult women in a society that has declared a "war on drugs." For women in communities devastated by poverty, this "war" has dramatically increased the penalties associated with what has evolved as both a coping strategy and a way to support themselves and their children. Far from the stereotypical drug kingpin, many low-level drug dealers and drug couriers are women attempting to make ends meet in communities where legitimate jobs are scarce to nonexistent and where even prostitution markets have collapsed.

The sixfold increase of the women's prison population since 1980 is an inadvertent but clear consequence of a society determined to crack down on crime, particularly on drug crimes. Haunted by increasing media images of amoral drug dealers and demonized strangers bent on vicious violence, the typical American is often quite surprised to discover that so many young women, many of whom are mothers, are being jailed because of the public's fears.

What else could we do? This book has identified a number of choices. Clearly, we could choose to decarcerate adult women as we decarcerated girls (particularly white girls) over the past 2 decades. The resulting release of adult women, like the earlier decarceration of girls, would be very unlikely to cause a surge in women's crime (particularly given the crimes for which women are serving time). Many, if not most, of the women being warehoused in U.S. prisons are in need of drug treatment and employment training. Those few who have been convicted of violent offenses have often killed an intimate (not infrequently someone who abused them) and are hardly likely to repeat the offense.

Reuniting women with their communities and their children is, at minimum, likely to save taxpayer money even if no additional treatment dollars are spent on programs to assist them (and their children) with their housing, employment, child care, and health needs. Because so many of the women sent to prison are guilty of no new crimes, dramatic decreases in prison populations could be achieved simply by policy changes aimed at reducing recommitment to prison for violations of probation and parole rules.

A society that, at one point in our history, had a vision of equality and social justice for all its citizens can surely better spend the great sums of money it is now costing us to incarcerate women offenders on improving their situation, and the situations of their children. Recall that over 70% of the women in prison have children (Bureau of Justice Statistics, 1999). For many of these children, their mother's incarceration signals a major trauma because their fathers rarely care for them. As a result, nearly three quarters of the children of female inmates are placed with relatives other than the natural father or in foster care, compared to only 10% of the children of male inmates (Donziger, 1996, p. 152). Because of the trauma associated with having a mother in prison, it makes sense that their children are far more likely than other children to end up in prison themselves (pp. 152–153).

Finally, as we think about the possibility of dramatically scaling back our reliance on imprisonment as a response to adult female offending, we might begin to consider such a response in the case of male offending. Why? The answer is simple. Currently, nearly one out of seven African American men age 25 to 29 are incarcerated, and overall, black males have a 29% chance of going to prison sometime during their lives (compared to 16% for Hispanic men and 4% for white men; Mauer, 1999; Sentencing Project, 2002b). As a result, our nation has the world's highest incarceration rate at 690 per 100,000 population, recently surpassing Russia's rate of 670 per 100,000 (Sentencing Project).

This book has focused on the consequences of the imprisonment binge for girls and women. Ending this book is impossible, however, without noting that this pattern has also signaled a dramatic increase in the imprisonment of the brothers, fathers, and sons of these women. As noted earlier in this book, crime has become a code word for race in the United States. As a result, correctional supervision, especially detention and imprisonment, seems increasingly to have replaced other historic systems of racial control (slavery, Jim Crow laws, ghettoization) as a way of keeping women and men of color in their "place" (Schiraldi, Kuyper, & Hewitt, 1996). This clearly has consequences for the girls and boys, women and men who are born nonwhite in a country with a lamentable history of racism. One scholar, commenting on this trend, observed, "'prison' is being re-lexified to become a code word for a terrible place where blacks reside" (Wideman, as cited in Schiraldi et al., 1996, p. 5).

The spiraling increase in the imprisonment of adult women is one of the most dramatic measures of this trend, but increases in the male prison population are also of great concern, particularly as they differentially affect ethnic communities in the United States. The cost to all of us for indulging in such unquestioning racism is only beginning to appear to the general public. The bill that the United States is currently paying for imprisonment is staggering. Currently, there are over 1.2 million sentenced prisoners in the United States. With an average daily operational expense of $30 per inmate, the cost of incarceration in the United States is now over $36 billion and climbing (see Department of Justice, 2001; Mauer, 1994). A conservative estimate is that each new prison cell costs about $100,000 to build and about $22,000 per bed to operate (Donziger, 1996, p. 49). As a direct result of the building boom in corrections, corrections budgets are by far the fastest growing segment of state budgets, increasing by 95% between 1976 and 1989. During this same period, state expenditures for lower education dropped slightly (2%), higher education dropped by 6%, and state expenditures for welfare (excluding Medicare) dropped by 41% (Donziger, p. 48). This means that money that once went to support low-income women and their children in the community, and to provide them with educational opportunities, is being cut back dramatically at the same time that money to arrest, detain, and incarcerate women on the economic margins is being increased.

If we are to respond to the challenge of girls' and women's crime, we must seek solutions that are based on the real causes of women's offenses, not on myths fostered by misinformation. We must understand how gender and race shape and eliminate choices for girls, how they injure (intentionally or not), and how they ultimately create very different futures for youths who are born female in a country that promises equality yet all too frequently falls short of that dream. We must also confront the fact that the United States has the highest rates of child poverty in the industrialized world (Donziger, 1996, p. 215),[1] and we must understand the ways in which this economic marginalization has directly affected girls and their mothers. Only with these understandings finally in mind can we imagine real solutions to the terrible problems of violence and crime in women's lives.

Again, as recent dramatic decreases in women's imprisonment in states such as California and New York demonstrate, we can do things differently. Moreover, in the case of California, it was the voters themselves who led the way. As state budgets face enormous deficits caused by the nation's economic woes, there is an even greater opportunity to get the public to understand that we must seek solutions to the nation's drug problems that do not involve the enormous human and economic cost associated with mass imprisonment (Mauer & Chesney-Lind, 2002).

We have seen how even the most perplexing of behaviors—girls' and women's violence—can be understood (but not excused) by the contexts that produced it. By listening to the voices of these girls and women and hearing their stories, we can also imagine other contexts and choices that, if provided, would allow them to do different things, to hope for a better future, and to be the people they are capable of becoming. Finally, we must understand that as we provide them with a brighter future, we guarantee a better future for ourselves, as well.

NOTE

1. About 46% of African American children and 39% of Hispanic children are born in poverty, compared to 16% of white children and 2% of children in Sweden. This last figure is particularly important because Sweden has a higher proportion of out-of-wedlock births than the United States (Donziger, 1996, p. 215).

REFERENCES

—·•·—

Acoca. L. (1999). Investing in girls: A 21st century challenge. *Juvenile Justice, 6*(1), 3-13.

Adams-Tucker, C. (1982). Proximate effects of sexual abuse in childhood. *American Journal of Psychiatry, 193,* 1252–1256.

Adler, F. (1975a). The rise of the female crook. *Psychology Today, 9,* 42–46, 112–114.

Adler, F. (1975b). *Sisters in crime.* New York: McGraw-Hill.

Adolescent Female Subcommittee. (1994). *Needs assessment and recommendations for adolescent females in Minnesota.* St. Paul, MN: Minnesota Department of Corrections.

Ageton, S. S. (1983). The dynamics of female delinquency, 1976–1980. *Criminology, 21,* 555–584.

Alder, C. (1986, December). "Unemployed women have got it heaps worse": Exploring the implications of female youth unemployment. *Australian and New Zealand Society of Criminology, 19,* 210–224.

Alder, C. (1995). *Delinquency prevention with young women.* Paper presented at the Delinquency Prevention Conference, Terrigal, Australia.

Alexander, R. (1995). *The "girl problem": Female sexual delinquency in New York, 1900–1930.* Ithaca, NY: Cornell University Press.

Amaro, H. (1995). Love, sex, and power: Considering women's realities in HIV prevention. *American Psychologist, 50,* 437–447.

Amaro, H., & Agular, M. (1994). *"Programa mama: Mom's project." A Hispanic/Latino family approach to substance abuse prevention.* Washington, DC: Department of Health and Human Services, Center for Substance Abuse Prevention, Mental Health Services Administration.

American Association of University Women. (1992). *How schools are shortchanging girls.* Washington, DC: American Association of University Women Educational Foundation.

American Bar Association and the National Bar Association. (2001). *Justice by gender: The lack of appropriate prevention, diversion and treatment alternatives for girls in the justice system.* Washington, DC: Author.

American Bar Foundation. (1995). Reducing crime by increasing incarceration: Does this policy make sense? *Researching Law, 6*(1), 1, 4–7.

177

American Correctional Association. (1990). *The female offender: What does the future hold?* Washington, DC: St. Mary's.

Anderson, S. (1994). *Comparison of male and female admissions one year prior to implementation of structured sanctions.* Salem, OR: Oregon Department of Corrections.

Andrews, R. H., & Cohn, A. H. (1974). Ungovernability: The unjustifiable jurisdiction. *Yale Law Journal, 83,* 1383–1409.

Aquino, B. (1994). *Filipino women and political engagement* (Working Paper Series Vol. 2). Honolulu, HI: The Office for Women's Research.

Armstrong, L. (1994). Who stole incest? *On the Issues, 3,* 30–32.

Arnold, R. (1995). The processes of victimization and criminalization of black women. In B. R. Price & N. Sokoloff (Eds.), *The criminal justice system and women* (pp. 136–146). New York: McGraw-Hill.

Austin, J., Bloom, B., & Donahue, T. (1992). *Female offenders in the community: An analysis of innovative strategies and programs.* Washington, DC: National Institute of Corrections, National Council on Crime and Delinquency.

Barnett, B. M. (1993). Invisible southern black women leaders in the civil rights movement: The triple constraints of gender, race, and class. *Gender and Society, 7,* 162–182.

Barry, K. (1996). Deconstructing deconstructionism (or whatever happened to feminist studies). In D. Bell & R. Klein (Eds.), *Radically speaking: Feminism reclaimed* (pp. 188–192). Melbourne, Australia: Spinifex.

Bartollas, C. (1993). Little girls grown up: The perils of institutionalization. In C. Culliver (Ed.), *Female criminality: The state of the art* (pp. 469–482). New York: Garland.

Baskin, D., & Sommers, I. (1993). Females' initiation into violent street crime. *Justice Quarterly, 10,* 559–581.

Baskin, D., Sommers, I., & Fagan, J. (1993). The political economy of female violent street crime. *Fordham Urban Law Journal, 20*(3), 401–417.

Baum, D. (1996). *Smoke and mirrors: The war on drugs and the politics of failure.* Boston: Little, Brown.

Beck, A. J., & Harrison, P. M. (2001). *Prisoners in 2000.* Washington, DC: Bureau of Justice Statistics.

Becker, H. S. (1963). *Outsiders.* New York: Free Press.

Beddoe, D. (1979). *Welsh convict women.* Barry, Wales: Stewart Williams.

Bell, I. P. (1970). The double standard: Age. *Transaction, 8,* 75–80.

Bergsmann, I. R. (1989). The forgotten few: Juvenile female offenders. *Federal Probation March 53*(1), 73–78.

Bishop, D., & Frazier, C. (1992). Gender bias in the juvenile justice system: Implications of the JJDP Act. *The Journal of Criminal Law and Criminology, 82,* 1162–1186.

Block, J. (1984). *Sex role identity and ego development.* San Francisco: Jossey-Bass.

Bloom, B., Chesney-Lind, M., & Owen, B. (1994). *Women in prison in California: Hidden victims of the war on drugs.* San Francisco: Center on Juvenile and Criminal Justice.

Bloom, B., & Steinhart, D. (1993). *Why punish the children?* San Francisco: National Council on Crime and Delinquency.

Blumstein, A., Cohen, J., Martin, S. E., & Tonry, M. H. (Eds.). (1983). *Research on sentencing: The search for reform* (Vols. 1–2). Washington, DC: National Academy Press.

Boritch, H., & Hagan, J. (1990). A century of crime in Toronto: Gender, class and patterns of social control, 1859–1955. *Criminology, 28,* 567–599.

Bourgois, P., & Dunlap, E. (1993). Exorcising sex-for-crack: An ethnographic perspective from Harlem. In M. Ratner (Ed.), *The crack pipe as pimp* (pp. 97–132). Lexington, MA: Lexington Books.

Bowker, L. (1978). *Women, crime and the criminal justice system.* Lexington, MA: Lexington Books.

Bowker, L., & Klein, M. (1983). The etiology of female juvenile delinquency and gang membership: A test of psychological and social structural explanations. *Adolescence, 13,* 739–751.

Brown, W. K. (1977). Black female gangs in Philadelphia. *International Journal of Offender Therapy and Comparative Criminology, 21,* 221–228.

Browne, A., & Finkelhor, D. (1986). Impact of child sexual abuse: A review of research. *Psychological Bulletin, 99,* 66–77.

Bureau of Justice Statistics. (1985). *Bulletin: Prisoners in 1984.* Washington, DC: U.S. Department of Justice.

Bureau of Justice Statistics. (1988). *Profile of state prison inmates, 1986.* Washington, DC: U.S. Department of Justice.

Bureau of Justice Statistics. (1989). *Criminal victimization in the United States.* Washington, DC: U.S. Department of Justice.

Bureau of Justice Statistics. (1993). *Jail inmates 1992.* Washington, DC: U.S. Department of Justice.

Bureau of Justice Statistics. (1999). *Women offenders.* Washington, DC: U.S. Department of Justice.

Bureau of Justice Statistics. (2001). *Prison and jail inmates at midyear 2001.* Washington, DC: U.S. Department of Justice.

Bureau of Justice Statistics. (2002a). *Prisoners in 2001.* Washington, DC: U.S. Department of Justice.

Bureau of Justice Statistics. (2002b). *Probation and parole in 2001.* Washington, DC: U.S. Department of Justice.

Burkhart, K. (1973). *Women in prison.* New York: Doubleday.

Bush-Baskette, S. R. (1999). The war on drugs: A war against women? In S. Cook & S. Davies (Eds.), *Harsh punishment* (pp. 211–229). Boston: Northeastern University Press.

Bynum, V. E. (1992). *Unruly women.* Chapel Hill: University of North Carolina Press.

Cain, M. (Ed.). (1989). *Growing up good: Policing the behavior of girls in Europe.* Newbury Park, CA: Sage.

Calahan, M. (1986). *Historical corrections statistics in the United States, 1850–1984.* Washington, DC: Bureau of Justice Statistics.

California Department of Justice. (1998). *1997 California criminal justice profile.* Sacramento, CA: Division of Criminal Justice Information Services.

California Department of Justice. (1999). *1998 California criminal justice profile.* Sacramento, CA: Division of Criminal Justice Information Services.

Cameron, M. B. (1953). *Department store shoplifting.* Unpublished doctoral dissertation, Indiana University.

Campagna, D. S., & Poffenberger, D. L. (1988). *The sexual trafficking in children.* Dover, MA: Auburn House.

Campbell, A. (1981). *Girl delinquents.* New York: St. Martin's.

Campbell, A. (1984). *The girls in the gang.* Oxford, UK: Basil Blackwell.

Campbell, A. (1990). Female participation in gangs. In R. Huff (Ed.), *Gangs in America* (pp. 163–182). Newbury Park, CA: Sage.

Campbell, A. (1993). *Men, women, and aggression.* New York: Basic Books.

Canter, R. J. (1982a). Family correlates of male and female delinquency. *Criminology, 20,* 149–167.

Canter, R. J. (1982b). Sex differences in self-report delinquency. *Criminology, 20,* 373–393.

Carmen, E., Crane, B., Dunnicliff, M., Holochuck, S., Prescott, L., Rieker, P., et al. (1996, January 25). *Massachusetts Department of Mental Health task force on the restraint and seclusion of persons who have been physically or sexually abused. Report and recommendations.* Boston: Massachusetts Department of Mental Health.

Carter, T. (1979). Juvenile court dispositions: A comparison of status and non-status offenders. *Criminology, 17,* 341–359.

CBS. (1992, August 6). Girls in the hood [Television series episode]. In *Street stories.*

Center for Policy Studies. (1991). *Violence against women as bias motivate hate crime.* Washington, DC: Center for Women Policy Studies.

Center for the Study of the States. (1993, November). State-local employment continues to grow. Albany, NY: *Rockefeller Institute of Government, 15,* 2.

Center on Juvenile and Criminal Justice. (2002). *About Proposition 21.* Retrieved December 15, 2002, from http://www.cjcj.org/jjic/prop_21.php#cp

Chain gang death. (1996, May 17). *Birmingham News,* p. 8A.

Chapman. J. R. (1980). *Economic realities and the female offender.* Lexington, MA: Lexington Books.

Chavkin, W. (1990). Drug addiction and pregnancy: Policy crossroads. *American Journal of Public Health, 80,* 483–487.

Chesney-Lind, M. (1971). *Female juvenile delinquency in Hawaii.* Unpublished master's thesis, University of Hawaii at Manoa.

Chesney-Lind, M. (1973). Judicial enforcement of the female sex role. *Issues in Criminology, 8,* 51–71.

Chesney-Lind, M. (1986). Women and crime: The female offender. *Signs, 12,* 78–96.

Chesney-Lind, M. (1987). Female offenders: Paternalism reexamined. In L. Crites & W. Hepperele (Eds.), *Women, the courts, and equality* (pp. 114–140). Newbury Park, CA: Sage.

Chesney-Lind, M. (1989). Girls' crime and woman's place: Toward a feminist model of female delinquency. *Crime and Delinquency, 35*, 5–30.

Chesney-Lind, M. (1993). Girls, gangs and violence: Reinventing the liberated female crook. *Humanity and Society, 17*, 321–344.

Chesney-Lind, M. (2002). The unintended victims of mass incarceration. In M. Chesney-Lind & M. Mauer (Eds.), *Invisible punishment: The collateral consequences of mass imprisonment* (pp. 79–94). New York: New Press.

Chesney-Lind, M., & Belknap, J. (2002, May). *Gender, delinquency, and juvenile justice: What about girls?* Paper presented at Aggression, Antisocial Behavior and Violence Among Girls: A Development Perspective: A Conference, Duke University, Durham, North Carolina.

Chesney-Lind, M., & Hagedorn, J. (1999). *Female gangs in America: Essays on girls, gangs, and gender.* Chicago: Lakeview Press.

Chesney-Lind, M., & Paramore, V. (1998, November). *Are girls getting more violent? Exploring juvenile robbery trends.* Paper presented at the annual meeting of the American Society of Criminology, Washington, DC.

Chesney-Lind, M., & Pasko, L. (2003, February). *Jailing girls.* Paper presented at the meeting of the Hawaii Sociological Association, Honolulu, Hawaii.

Chesney-Lind, M., & Pollock-Byrne, J. (1995). Women's prisons: Equality with a vengeance. In J. Pollock-Byrne & A. Merlo (Eds.), *Women, law, and social control* (pp. 155–175). Boston: Allyn & Bacon.

Chesney-Lind, M., Rockhill, A., Marker, N., & Reyes, H. (1994). Gangs and delinquency: Exploring police estimates of gang membership. *Crime, Law and Social Change, 21*, 201–228.

Chesney-Lind, M., & Rodriguez, N. (1983). Women under lock and key. *The Prison Journal, 63*, 47–65.

Chesney-Lind, M., Shelden, M. R., & Laidler, K. J. (1996). Girls, delinquency and gang membership. In R. Huff (Ed.), *Gangs in America* (2nd ed.). Thousand Oaks, CA: Sage.

Chesney-Lind, M., & Shelden, R. G. (1998). *Girls, delinquency, and the juvenile justice system.* Pacific Grove, CA: Brooks/Cole.

Chilton, R., & Datesman, S. K. (1987). Gender, race and crime: An analysis of urban arrest trends, 1960–1980. *Gender and Society, 1*, 152–171.

Chinen, J. (1994). Internationalization of capital, migration, reindustrialization and women workers in the garment industry. *Social Process in Hawaii, 35*, 85–102.

Clarke, S. H., & Koch, G. C. (1980). Juvenile court: Therapy and crime control, and do lawyers make a difference? *Law and Society Review, 14*, 263–308.

Cloward, R. A., & Ohlin, L. E. (1960). *Delinquency and opportunity.* New York: Free Press.

Cohen, A. K. (1955). *Delinquency in boys: The culture of the gang.* New York: Free Press.

Cohen, L. E., & Kluegel, J. R. (1979). Selecting delinquents for adjudication. *Journal of Research on Crime and Delinquency, 16,* 143–163.

Cohn, Y. (1970). Criteria for the probation officer's recommendation to the juvenile court. In P. G. Garbedian & D. C. Gibbons (Eds.), *Becoming delinquent* (pp. 262–275). Chicago: Aldine.

Coles, F. (1991, February). *Women, alcohol, and automobiles: A deadly cocktail.* Paper presented at the Western Society of Criminology Meetings, Berkeley, CA.

Conly, C. (1998). *The Women Prison's Association: Supporting women offenders and their families.* Washington, DC: National Institute of Justice.

Connell, R. W. (1987). *Gender and power.* Stanford, CA: Stanford University Press.

Corrado, R., Odgers, C., & Cohen, I. (2000, April). The incarceration of female young offenders: Protection for whom? *Canadian Journal of Criminology,* 189–207.

Costello, J. C., & Worthington, N. L. (1981). Incarcerating status offenders: Attempts to circumvent the Juvenile Justice and Delinquency Prevention Act. *Harvard Civil Rights—Civil Liberties Law Review, 16,* 41–81.

Craig, G. (1995, April 8). Videotaped frisks anger women inmates. *Rochester Democrat and Chronicle,* pp. 1A, 8A.

Craig, G. (1996, March 23). Advocates say nude filming shows need for new laws. *Rochester Democrat and Chronicle,* pp. A1, A6.

Crime down, media coverage up. (1994, January/February). *Media Monitor, 8*(1).

Crites, L. (1976). *The female offender.* Lexington, MA: Lexington Books.

Crittenden, D. (1990, January 25). You've come a long way, Moll. *Wall Street Journal,* p. A14.

Curriden, M. (1993, September 20). Prison scandal in Georgia: Guards traded favors for sex. *National Law Journal,* 8.

Curry, G. D. (1995, November). *Responding to female gang involvement.* Paper presented at the American Society of Criminology Meetings, Boston.

Curry, G. D., Ball, R. A., & Fox, R. J. (1994). *Gang crime and law enforcement record-keeping.* Washington, DC: National Institute of Justice.

Curry, G. D., Fox, R. J., Ball, R. A., & Stone, D. (1992). *National assessment of law enforcement anti-gang information resources: Draft 1992 final report* (National Assessment Survey 1992). Morgantown: West Virginia University.

Daly, K. (1989). Gender and varieties of white-collar crime. *Criminology, 27,* 769–793.

Daly, K., & Bordt, R. (1991). *Gender, race, and discrimination research: Disparate meanings of statistical "sex" and "race effects" in sentencing.* Department of Sociology: University of Michigan.

Daly, K. (1994). *Gender, crime, and punishment.* New Haven, CT: Yale University Press.

Daly, K., & Chesney-Lind, M. (1988). Feminism and criminology. *Justice Quarterly, 5,* 497–538.

Datesman, S., & Scarpitti, F. (1977). Unequal protection for males and females in the juvenile court. In T. N. Ferdinand (Ed.), *Juvenile delinquency: Little brother grows up* (pp. 59–77). Beverly Hills, CA: Sage.

Davidson, S. (Ed.). (1983). *The second mile: Contemporary approaches in counseling young women.* Tucson, AZ: New Directions for Young Women.

DeJong, A. R., Hervada, A. R., & Emmett, G. A. (1983). Epidemiologic variations in childhood sexual abuse. *Child Abuse and Neglect, 7,* 155–162.

Dembo, J. S., Sue, C. C., Borden, P., & Manning, D. (1995, August). *Gender differences in service needs among youths entering a juvenile assessment center: A replication study.* Paper presented at the annual meeting of the Society of Social Problems, Washington, DC.

Dembo, R., Williams, L., & Schmeidler, J. (1993). Gender differences in mental health service needs among youths entering a juvenile detention center. *Journal of Prison and Jail Health, 12,* 73–101.

Department of Justice. (2001). *Emerging issues on privatized prisons.* Washington, DC: Author.

Deschenes, E., & Anglin, D. M. (1992). Effects of legal supervision on narcotic addict behavior: Ethnic and gender influences. In T. Mieczkowski (Ed.), *Drugs, crime, and social policy* (pp. 167–196). Needham, MA: Allyn & Bacon.

Deschenes, E., & Esbensen, F. (1999). Violence among girls: Does gang membership make a difference?. In M. Chesney-Lind & J. Hagedorn (Eds.), *Female Gangs in America.* Chicago: Lake View Press.

Doi, D. (2002, October 10). Reauthorization update. *Memorandum to Coalition for Juvenile Justice.* Washington, DC: Coalition for Juvenile Justice.

Donziger, S. (Ed.). (1996). *The real war on crime.* New York: HarperPerennial.

Dungworth, T. (1977). Discretion in the juvenile justice system. In T. N. Ferdinand (Ed.), *Juvenile delinquency: Little brother grows up* (pp. 19–44). Beverly Hills, CA: Sage.

Eaton, M. (1986). *Justice for women?* Milton Keynes, UK: Open University Press.

Elis, L., MacKenzie, D., & Simpson, S. (1992, October). *Women and shock incarceration.* Paper presented at the Focus group meeting, Department of Criminology, College Park, MD: University of Maryland.

English, K. (1993). Self-reported crimes rates of women prisoners. *Journal of Quantitative Criminology, 9,* 357–382.

Enloe, C. (1989). *Bananas, beaches and bases: Making feminist sense of international politics.* Berkeley: University of California Press.

Enos, S. (2001). *Mothering from the inside: Parenting in a women's prison.* New York: State University of New York Press.

Erickson, P., Butters, J., McGillicuddy, P., & Hallgren, A. (2000). Crack and prostitution: Gender, myths, and experiences. *Journal of Drug Issues, 30,* 767–789.

Faludi, S. (1991). *Backlash: The undeclared war against women.* New York: Crown.

Federal Bureau of Investigation. (1973). *Crime in the United States—1972.* Washington, DC: U.S. Department of Justice.

Federal Bureau of Investigation. (1976). *Crime in the United States—1975.* Washington, DC: U.S. Department of Justice.

Federal Bureau of Investigation. (1980). *Crime in the United States—1979: Uniform crime reports.* Washington, DC: U.S. Department of Justice.

Federal Bureau of Investigation. (1994). *Crime in the United States—1993*. Washington, DC: U.S. Department of Justice.

Federal Bureau of Investigation. (1995). *Crime in the United States—1994*. Washington, DC: U.S. Department of Justice.

Federal Bureau of Investigation. (2002). *Crime in the United States—2001*. Washington, DC: U.S. Department of Justice.

Feeley, M., & Little, D. L. (1991). The vanishing female: The decline of women in the criminal process. *Law and Society Review, 256*, 719–758.

Feinman, C. (1980). *Women in the criminal justice system*. New York: Praeger.

Female Offender Resource Center. (1977). *Little sisters and the law*. Washington, DC: American Bar Association.

Fessendon, F. (2000, April 9). They threaten, seethe, and unhinge, then kill in quantity. *New York Times*, sec. 1, p. 1.

Figueira-McDonough, J. (1985). Are girls different? Gender discrepancies between delinquent behavior and control. *Child Welfare, 64*, 273–289.

Figueira-McDonough, J., & Selo, E. (1980). A reformulation of the "equal opportunity" explanation of female delinquency. *Crime and Delinquency, 26*, 333–343.

Finkelhor, D. (1982). Sexual abuse: A sociological perspective. *Child Abuse and Neglect, 6*, 95–102.

Finkelhor, D., & Baron, L. (1986). Risk factors for child sexual abuse. *Journal of Interpersonal Violence, 1*, 43–71.

Fishman, L. T. (1995). The Vice Queens: An ethnographic study of black female gang behavior. In M. Klein, C. Maxson, & J. Miller (Eds.), *The modern gang reader* (pp. 83–92). Los Angeles: Roxbury.

Flowers, R. B. (1986). *Children and criminality: The child as victim and perpetrator*. Westport, CT: Greenwood.

Flowers, R. B. (1987). *Women and criminality*. Westport, CT: Greenwood.

Flowers, R. B. (2001). *Runaway kids and teenage prostitution*. London: Greenwood.

Foley, C. (1974, October 20). Increase of women in crime and violence. *Honolulu Sunday Star-Bulletin and Advertiser*, p. 1.

Franklin, R. (1996, April 26). Ala. to expand chain gangs—adding women. *USA Today*, p. 3A.

Freedman, E. (1981). *Their sisters' keepers*. Ann Arbor: University of Michigan Press.

Fullilove, M., Lown, A., & Fullilove, R. (1992). Crack hos and skeezers: Traumatic experiences of women crack users. *The Journal of Sex Research, 29*, 275–287.

Gelsthorpe, L. (1989). *Sexism and the female offender: An organizational analysis*. Aldershot, UK: Gower.

General Accounting Office. (1978). *Removing status offenders from secure facilities: Federal leadership and guidance are needed*. Washington, DC: Author.

Gibbons, D. (1983). *Delinquent behavior*. Englewood Cliffs, NJ: Prentice Hall.

Gibbons, D., & Griswold, M. J. (1957). Sex differences among juvenile court referrals. *Sociology and Social Research, 42*, 106–110.

Gibbs, B. (June 19, 2001). *Number of girls in gangs increasing*. Retrieved on December 15, 2002, from http://abclocal.go.com/wtvd/features/061901_CF_girlsgangs.html

Gilfus, M. (1992). From victims to survivors to offenders: Women's routes of entry into street crime. *Women and Criminal Justice, 4*(1), 63–89.

Giordano, P., Cernkovich, S., & Pugh, M. (1978). Girls, guys and gangs: The changing social context of female delinquency. *Journal of Criminal Law and Criminology, 69*, 126–132.

Girls Incorporated. (1996). *Prevention and parity: Girls in juvenile justice.* Indianapolis, IN: Girls Incorporated National Resource Center.

Girshick, L. (1999). *No safe haven: Stories of women in prison*. Boston: Northeastern University Press.

Goldberg, L. (1996, February 16). Juvenile hall strip search of girls spurs questions. *San Francisco Examiner*, p. A1.

Gora, J. (1982). *The new female criminal: Empirical reality or social myth*. New York: Praeger.

Gordon, L. (1988). *Heroes in their own lives*. New York: Viking.

Green, P. (Ed.). (1996). *Drug couriers: A new perspective*. London: Quartet.

Greene, Peters, and Associates. (1998). *Guiding principles for promising female programming: An inventory of best practices.* Nashville, TN: Office of Juvenile Justice and Delinquency Prevention.

Hagan, J., Gillis, A. R., & Simpson, J. (1985). The class structure of gender and delinquency: Toward a power-control theory of common delinquent behavior. *American Journal of Sociology, 90*, 1151–1178.

Hagan, J., Simpson, J., & Gillis, A. R. (1987). Class in the household: A power-control theory of gender and delinquency. *American Journal of Sociology, 92*, 788–816.

Hanawalt, L. B. (1982). Women before the law: Females as felons and prey in 14th-century England. In D. K. Weisberg (Ed.), *Women and the law* (pp. 165–196). Cambridge, MA: Schenkman.

Hancock, L. (1981). The myth that females are treated more leniently than males in the juvenile justice system. *Australian and New Zealand Journal of Sociology, 16*, 4–14.

Hancock, L., & Chesney-Lind, M. (1982). Female status offenders and justice reforms: An international perspective. *Australian and New Zealand Journal of Criminology, 15*, 109–122.

Hanson, K. (1964). *Rebels in the streets: The story of New York's girl gangs*. Englewood Cliffs, NJ: Prentice Hall.

Harlow, C. W. (1999). *Prior abuse reported by inmates and probationers*. Washington, DC: U.S. Department of Justice.

Harms, P. (2002). *Detention in delinquency cases, 1989-1998* (OJJDP fact sheet, No. 1). Washington, DC: U.S. Department of Justice.

Harris, M. G. (1988). *Cholas: Latino girls and gangs*. New York: AMS Press.

Hartman, M. S. (1977). *Victorian murderesses*. New York: Schocken.

Herman, J. L. (1981). *Father-daughter incest*. Cambridge, MA: Harvard University Press.

Hippensteele, S., & Chesney-Lind, M. (1995). Race and sex discrimination in the academy. *Thought and Action, 11*(2), 43–66.

Hirschi, T. (1969). *Causes of delinquency*. Berkeley: University of California Press.

Hochschild, A. (1989). *The second shift*. New York: Viking.

Holsinger, K., Belknap, J., & Sutherland, J. (1999). *Assessing the gender specific program and service needs for adolescent females in the juvenile justice system*. Columbus, OH: Office of Criminal Justice Services.

Howard, B. (1996, July/August). Juvenile Justice Act's mandates: Stay or go? *Youth Today*, 22.

Hser, Y., Anglin, M. D., & Chou, C. (1992). Narcotics use and crime among addicted women: Longitudinal patterns and effects of social interventions. In T. Mieczkowski (Ed.), *Drugs, crime, and social policy* (pp. 197–221). Needham, MA: Allyn & Bacon.

Huff, R. (Ed.). (2002). *Gangs in America* (3rd ed.). Newbury Park, CA: Sage.

Hulen, T. (1996, April 28). Governor's stand on women in chains: Insult or chivalry. *Birmingham News*, pp. 1A, 2A.

Huling, T. (1995, November). *African American women and the war on drugs*. Paper presented at the annual meeting of the American Society of Criminology, Boston.

Huling, T. (1996). Prisoners of war: Women drug couriers in the United States. In P. Green (Ed.), *Drug couriers: A new perspective* (pp. 46–60). London: Quartet.

Human Rights Watch. (1993). *The Human Rights Watch global report on prisons*. New York: Author.

Hunt, G., & Joe-Laidler, K. (2001). Situations of violence in the lives of girl gang members. *Health Care for Women International, 22*, 363–384.

Ianni, F.A.J. (1989). *The search for structure: A report on American youth today*. New York: Free Press.

Inciardi, J., Lockwood, D., & Pottieger, A. E. (1993). *Women and crack-cocaine*. New York: Macmillan.

Iwamoto, J. J., Kameoka, K., & Brasseur, Y. C. (1990). *Waikiki homeless youth project: A report*. Honolulu: Catholic Services to Families.

James, J. (1976). Motivations for entrance into prostitution. In L. Crites (Ed.), *The female offender* (pp. 177–206). Lexington, MA: Lexington Books.

Jamieson, K. M., & Flanagan, T. (Eds.). (1989). *Sourcebook of criminal justice statistics—1988*. Washington, DC: U.S. Department of Justice, Bureau of Justice Statistics.

Jankowski, M. S. (1991). *Islands in the streets: Gangs and American urban society*. Berkeley: University of California Press.

Joe, K. (1993). Getting into the gang: Methodological issues in studying ethnic gangs. In M. De La Rosa & J. L. Adrados (Eds.), *Drug abuse among minority youth: Advances in research and methodology* (pp. 234–237). National Institute on Drug Abuse Research Monograph No. 130.

Joe, K. (1995a, April). *Asian Pacific American women drug users: An ethnographic study*. Paper presented at the annual meetings of the Pacific Sociological Association, San Francisco.

Joe, K. (1995b). Ice is strong enough for a man but made for a woman: A social cultural analysis of methamphetamine use among Asian Pacific Americans. *Crime, Law and Social Change, 22*, 269–289.

Joe, K. (1996a). *Going home: The double edged sword*. Unpublished manuscript.

Joe, K. (1996b). The life and times of Asian American women drug users: An ethnographic study. *Journal of Drug Issues, 26*(1), 125–142.

Joe, K., & Chesney-Lind, M. (1995). Just every mother's angel: An analysis of gender and ethnic variations in youth gang membership. *Gender and Society, 9*, 408–430.

Joe, K., & Morgan, P. (1997). Kinship and community: The ice crisis in Hawaii. In H. Klee (Ed.), *Amphetamine misuse: International perspectives on current trends* (pp. 163-180). London: Gordon and Breach.

Johnson, D. R., & Scheuble, L. K. (1991). Gender bias in the disposition of juvenile court referrals: The effects of time and location. *Criminology, 29*(4), 677–699.

Jones, A. (1980). *Women who kill*. New York: Fawcett.

Kahler, K. (1992, May 17). Hand that rocks the cradle is taking up violent crime. *Sunday Star-Ledger*, p. 3A.

Kamler, B. (1999). *Constructing gender and difference: Critical perspectives on early childhood*. Cresskill, NJ: Hampton Press.

Katz, P. A. (1979). The development of female identity. In C. B. Kopp (Ed.), *Becoming female: Perspectives on development* (pp. 3–27). New York: Plenum.

Kauffman, H. M. (1993). *In search of the "new violent female offender" in Hawaii*. Unpublished honor's thesis, University of Hawaii at Manoa.

Kim, E. (1996, August 16). Sheriff says he'll have chain gangs for women. *Tuscaloosa News*, p. 1A.

Kirp, D., Yudof, M., & Franks, M. S. (1986). *Gender justice*. Chicago: University of Chicago Press.

Kirwan, S. (2000, December 5). Youth facility set for discussion. *The Los Angeles Times*. Retrieved December 15, 2002, from www.latimes.com.

Klein, D., & Kress, J. (1976, Spring/Summer). Any woman's blues: A critical overview of women, crime and the criminal justice system. *Crime and Social Justice, 5*, 34–48.

Klein, M. (2002). Street gangs: A cross national perspective. In C.R. Huff (Ed.), *Gangs in America* (3rd ed., pp. 237–254). Thousand Oaks, CA: Sage.

Klemesrud, J. (1978, January 16). Women terrorists, sisters in crime. *Honolulu Star Bulletin*, p. C1.

Koop, C. E. (1989, May 22). *Violence against women: A global problem*. Address by the Surgeon General of the U.S. Public Health Service at a conference of the Pan American Health Organization, Washington, DC.

Kratcoski, P. C. (1974). Delinquent boys and girls. *Child Welfare, 5*, 16–21.

Krisberg, B., DeComo, R., Herrera, N. C., Steketee, M., & Roberts, M. (1991).
    Juveniles taken into custody: Fiscal year 1990 report. San Francisco: National
    Council on Crime and Delinquency.

Krisberg, B., Schwartz, I. M., Fishman, G., Eisikovits, Z., & Guttman, E. (1986). *The
    incarceration of minority youth*. Minneapolis, MN: Hubert Humphrey Institute of
    Public Affairs.

Kunzel, R. (1993). *Fallen women and problem girls: Unmarried mothers and the
    professionalization of social work, 1890–1945*. New Haven, CT: Yale University
    Press.

LaFromboise, T. D., & Howard-Pitney, B. (1995). Suicidal behavior in American
    Indian female adolescents. In S. Canetto & D. Lester (Eds.), *Woman and suicidal
    behavior* (pp. 157–173). New York: Springer.

Langan, P. A. (1991, March 29). America's soaring prison population. *Science, 251*,
    1569.

Lauderback, D., Hansen, J., & Waldorf, D. (1992). Sisters are doin' it for themselves:
    A black female gang in San Francisco. *The Gang Journal, 1*, 57–72.

LeBlanc, A. (1995). Trina and Trina. *Literary Journalism, 2*, 213–233.

LeBlanc, A. (1996, June 2). A woman behind bars is not a dangerous man. *New York
    Times Magazine*, 34–40.

Lee, F. R. (1991, November 25). For gold earrings and protection, more girls take the
    road to violence. *New York Times*, pp. A1, B7.

Leslie, C., Biddle, N., Rosenberg, D., & Wayne, J. (1993, August 2). Girls will be girls.
    Newsweek, 44.

Lewis, N. (1992, December 23). Delinquent girls achieving a violent equality in DC.
    The Washington Post, pp. A1, A14.

Lindquist, J. (1988). *Misdemeanor crime*. Newbury Park, CA: Sage.

Linnekin, J. (1990). *Sacred queens and women of consequence*. Ann Arbor: University
    of Michigan Press.

Lipsey, M. (1992). Juvenile delinquency treatment: A meta-analytic inquiry in the
    variability of effects. In T. A. Cook, H. Cooper, D. S. Cordray, H. Hartmann,
    L. V. Hedges, R. J. Light, et al. (Eds.), *Meta-analysis for explanation: A casebook*
    (pp. 83–126). New York: Russell Sage.

Loper, A. B., & Cornell, D. G. (1996). *Homicide by girls*. Paper presented at the annual
    meeting of the National Girls Caucus, Orlando, FL.

Lopez, S. (1993, July 8). Fifth guard arrested on sex charge. *Albuquerque Journal*,
    pp. A1, A2.

Los Angeles Times Service. (1975, August 7). L.A. police chief blames libbers.
    Honolulu Advertiser, pp. B4.

MacKinnon, C. (1987). *Feminism unmodified: Discourses on life and law*. London:
    Harvard University Press.

Maher, L., & Curtis, R. (1992). Women on the edge: Crack cocaine and the changing
    contexts of street-level sex work in New York City. *Crime, Law and Social
    Change, 18*, 221–258.

Maher, L., Dulap, E., Johnson, B., & Hamid, A. (1996). Gender, power, and alternative living arrangements in the inner city crack culture. *Journal of Research in Crime and Delinquency, 33*, 181–205.

Males, M. (1994, March/April). Bashing youth: Media myths about teenagers. *Extra*, 8–11.

Mann, C. (1979). The differential treatment between runaway boys and girls in juvenile court. *Juvenile and Family Court Journal, 30*, 37–48.

Mann, C. (1984). *Female crime and delinquency*. Tuscaloosa: University of Alabama Press.

Maquire, K., & Pastore, A. L. (Eds.). (1994). *Sourcebook of criminal justice statistics— 1993*. Washington, DC: U.S. Department of Justice.

Martin, M. (2002, April 21). Changing population behind bars: Major drop in women in state prisons. *San Francisco Chronicle*.

Maruschak, L. (2001). *HIV in prisons and jails, 1999*. Washington, DC: Bureau of Justice Statistics.

Maruschak, L. (2002). *HIV in prisons, 2000*. Washington, DC: Bureau of Justice Statistics.

Mauer, M. (1994, September). *Americans behind bars: The international use of incarceration, 1992–1993*. Washington, DC: The Sentencing Project.

Mauer, M. (1999). *The crisis of the young African American male and the criminal justice system*. Washington, DC: The Sentencing Project.

Mauer, M., & Chesney-Lind, M. (2002). *Invisible punishment: The collateral consequences of mass imprisonment*. New York: The New Press.

Mauer, M., & Huling, T. (1995). *Young Black Americans and the criminal justice system: Five years later*. Washington, DC: The Sentencing Project.

May, D. (1977). Delinquent girls before the courts. *Medical Science Law Review, 17*, 203–210.

Mayer, J. (1994, July). *Girls in the Maryland juvenile justice system: Findings of the Female Population Taskforce*. Paper presented at the Gender Specific Services Training, Minneapolis, MN.

McClellan, D. S. (1994). Disparity in the discipline of male and female inmates in Texas prisons. *Women and Criminal Justice, 5*(2), 71–97.

McCormack, A., Janus, M., & Burgess, A. W. (1986). Runaway youths and sexual victimization: Gender differences in an adolescent runaway population. *Child Abuse and Neglect, 10*, 387–395.

McDermott, M. J., & Blackstone, S. J. (1994). *White slavery plays of the 1910's: Fear of victimization and the social control of sexuality*. Paper presented at the annual meeting of the American Society of Criminology, Miami, FL.

McRobbie, A., & Garber, J. (1975). Girls and subcultures. In S. Hall & T. Jefferson (Eds.), *Resistance through rituals: Youth subculture in post-war Britain* (pp. 209–222). New York: Holmes and Meier.

Meiselman, K. (1978). *Incest*. San Francisco: Jossey-Bass.

Merton, R. K. (1938). Social structure and anomie. *American Sociological Review, 3*, 672–682.

Messerschmidt, J. (1987). Feminism, criminology, and the rise of the female sex delinquent, 1880–1930. *Contemporary Crises, 11*, 243–263.

Meyer, M. (1992, November 9). Coercing sex behind bars: Hawaii's prison scandal. *Newsweek*, 23–25.

Miller, E. (1986). *Street woman*. Philadelphia: Temple University Press.

Miller, J. (1994). Race, gender and juvenile justice: An examination of disposition decision-making for delinquent girls. In M. D. Schwartz & D. Milovanovic (Eds.), *The intersection of race, gender and class in criminology* (pp. 219–246). New York: Garland.

Miller, J. (2001). *One of the guys: Girls, gangs, and gender*. New York: Oxford University Press.

Miller, W. B. (1958). Lower class culture as a generating milieu of gang delinquency. *Journal of Social Issues, 14*, 5–19.

Miller, W. B. (1975). *Violence by youth gangs and youth groups as a crime problem in major American cities*. Washington, DC: Government Printing Office.

Miller, W. B. (1980). The Molls. In S. K. Datesman & F. R. Scarpitti (Eds.), *Women, crime, and justice* (pp. 238–248). New York: Oxford University Press.

Moone, J. (1993a). *Children in custody: Private facilities*. Washington, DC: Office of Juvenile Justice and Delinquency Prevention.

Moone, J. (1993b). *Children in custody: Public facilities*. Washington, DC: Office of Juvenile Justice and Delinquency Prevention.

Moore, J. (1991). *Going down to the barrio: Homeboys and homegirls in change*. Philadelphia: Temple University Press.

Moore, J., & Hagedorn, J. (1995). What happens to the girls in the gang? In R. C. Huff (Ed.), *Gangs in America* (2nd. ed., pp. 205–220). Thousand Oaks, CA: Sage.

Moore, J., Vigil, D., & Levy, J. (1995). Huisas of the street: Chicana gang members. *Latino Studies Journal, 6*(1), 27–48.

Morash, M., & Bynum, T. (1996). *Findings from the national study of innovative and promising programs for women offenders*. East Lansing: Michigan State University, School of Criminal Justice.

Morgan, E. (2000). Women on death row. In R. Muraskin (Ed.), *It's a crime: Women and justice* (pp. 269–283). Upper Saddle River, NJ: Prentice Hall.

Morgan, P., Beck, J., Joe, K., McDonnell, D., & Gutierrez, R. (1994). *Report of findings. Ice and other methamphetamine use: An exploratory study. Final report to the National Institute on Drug Abuse*. San Francisco: National Institute on Drug Abuse, Institute for Scientific Analysis.

Morgan, P., & Joe, K. (1996). Citizens and outlaws: The private lives and public lifestyles of women in the illicit drug economy. *Journal of Drug Issues, 26*(1), 199–218.

Morgan, P., & Joe, K. (1997). Uncharted terrains: Contexts of experience among women in the illicit drug economy. *Women and Criminal Justice, 8*(3), 85–110.

Morris, A. (1987). *Women, crime and criminal justice*. New York: Basil Blackwell.

Muwakkil, S. (1993, April 5). Ganging together. *In These Times*, pp. 14–18.

Naffine, N. (1987). *Female crime: The construction of women in criminology.* Sydney, Australia: Allen and Unwin.

Naffine, N. (1989). Toward justice for girls. *Women and Criminal Justice, 1*, 3–19.

National Institute of Justice. (1998). *Women offenders: Programming needs and promising approaches.* Washington, DC: Author.

National Symposium on Female Offenders. (2000). *Conference proceedings: Treat the women, save the children.* Kauai, HI: Author.

NBC. (1993, March 29). *NBC nightly news.* [Television broadcast]. Diana Koricke in East Los Angeles.

Nelson, L. D. (1977, October 23). Women make gains in shady world, too. *Honolulu Sunday Star-Bulletin and Advertiser*, p. G-8.

Noble, A. (1988). *Criminalize or medicalize: Social and political definitions of the problem of substance use during pregnancy.* Sacramento, CA: Department of Health Services, Maternal and Child Health Branch.

Nunes, K., & Whitney, S. (1994, July). The destruction of the Hawaiian male. *Honolulu Magazine, 29, 1*, 58–60.

Odem, M. E., & Schlossman, S. (1991). Guardians of virtue: The juvenile court and female delinquency in early 20th century Los Angeles. *Crime and Delinquency, 37*, 186–203.

Office of Juvenile Justice and Delinquency Prevention. (1992). *Arrests of youth 1990.* Washington, DC: Author.

Office of Juvenile Justice and Delinquency Prevention. (1998). *Guiding principles for promising female programming.* Washington, DC: Author.

Office of Juvenile Justice and Delinquency Prevention. (2001). *OJJDP statistical briefing book.* Retrieved December 15, 2002, from http: //www.ojjdp.ncjrs.org/ojstabb/qa178.html

Okamura, J. (1990). *Ethnicity and stratification in Hawaii.* (Operation Manong Resource Papers No. 1.) Honolulu: University of Hawaii at Manoa, Operation Manong Program.

Okamura, J. (1994). Why there are no Asian Americans in Hawaii: The continuing significance of local identity. *Social Process in Hawaii, 35*, 161–178.

Orenstein, P. (1994). *School Girls.* Garden City, NY: Doubleday.

Ostner, I. (1986). Die Entdeckung der Mädchen. Neue Perspecktiven für die. *Kolner-Zeitschrift für Soziologie und Sozialpsychologie, 38*, 352–371.

O'Toole, L., & Schiffman, J. (1997). *Gender violence: Interdisciplinary perspectives.* New York: New York University Press.

Ouellet, L. J., Wiebel, W. W., Jimenez, A. D., & Johnson, W. A. (1993). Crack cocaine and the transformation of prostitution in three Chicago neighborhoods. In M. Ratner (Ed.), *The crack pipe as pimp* (pp. 69–95). New York: Lexington.

Owen, B. (1998). *In the mix: Struggle and survival in a women's prison.* New York: State University of New York.

Parham v. J. R., 442 U.S. 584 (1979).

Pasko, L. (1997). *From sin to syndrome: The medicalization of juvenile sex offense.* Unpublished master's thesis, University of Nevada, Reno.

Patrick, D. L. (1995, March 27). Letter to Gov. John Engler, Re: Crane and Scott Correctional Centers.

Phelps, R. J., McIntosh, M., Jesudason, V., Warner, P., & Pohlkamp, J. (1982). *Wisconsin female juvenile offender study project summary report.* Madison, WI: Wisconsin Council on Juvenile Justice, Youth Policy and Law Center.

Platt, A. M. (1969). *The childsavers.* Chicago: University of Chicago Press.

Poe, E., & Butts, J. A. (1995). *Female offenders in the juvenile justice system.* Pittsburgh, PA: National Center for Juvenile Justice.

Poe-Yamagata, E., & Butts, J. A. (1996). *Female offenders in the juvenile justice system.* Washington, DC: U.S. Department of Justice.

Pohl, J., & Boyd, C. (1992). Female addiction: A concept analysis. In T. Mieczkowski (Ed.), *Drugs, crime, and social policy* (pp. 138–152). Needham, MA: Allyn & Bacon.

Pollock-Byrne, J. (1990). *Women, prison, and crime.* Pacific Grove, CA: Brooks/Cole.

Pope, C., & Feyerherm, W. H. (1982). Gender bias in juvenile court dispositions. *Social Service Research, 6,* 1–17.

Portillos, E., & Zatz, M. S. (1995, November). *Not to die for: Positive and negative aspects of Chicano youth gangs.* Paper presented at the meeting of the American Society of Criminology, Boston.

Province of British Columbia. (1978). *Youth services in juvenile justice.* Victoria, Canada: Information Services, Corrections Branch.

Public Law 102–586—(November 4, 1992). Juvenile justice and delinquency prevention, fiscal years 1993, 1994, 1995, 1996. 106 Stat. 4982, (1992).

Pukui, M. L., Haertig, E. W., & Lee, C. (1972). *Nana I Ke Kumu* (Vol. 1). Honolulu, HI: Hui Hanai.

Puzzanchera, C., Stahl, A. L., Finnegan, T. A., Snyder, H., Poole, R. S., & Tierney, N. (2000). *Juvenile court statistics 1997.* Washington, DC: National Center for Juvenile Justice.

Pyett, P. M., & Warr, D. J. (1997). Vulnerability on the streets: Female sex workers and HIV risk. *AIDS Care, 9,* 539–547.

Pyett, P. M., & Warr, D. J. (1999). Women at risk in sex work: Strategies for survival. *Journal of Sociology, 35*(2), 183–197.

Quicker, J. C. (1983). *Homegirls: Characterizing Chicano gangs.* San Pedro, CA: International University Press.

Raeder, M. (1993). *Gender and sentencing: Single moms, battered women and other sex-based anomalies in the gender free world of the federal sentencing guidelines.* Unpublished manuscript.

Rafter, N. H. (1990). *Partial justice: Women, prisons and social control.* New Brunswick, NJ: Transaction Books.

Rans, L. (1975). *Women's arrest statistics. (The women offender report).* Washington, DC: American Bar Association, Female Offender Resource Center.

Reitsma-Street, M. (1993). Canadian youth court charges and dispositions for females before and after implementation of the Young Offenders Act. *Canadian Journal of Criminology, 35,* 437–458.

Rennison, C. M. (2001). *Intimate and partner violence and age of victim*. Washington, DC: Bureau of Justice Statistics.

Renzetti, C., & Curran, D. J. (1995). *Women, men, and society*. Boston: Allyn & Bacon.

Rice, R. (1963, October 19). A reporter at large: The Persian Queens. *New Yorker*, 153–187.

Rimbach, J. (1994). Bad prison: Tools at Skillman: Medication, isolation. In *Delinquent justice: A special reprint* (pp. 21–40). Haekensack, New Jersey: The Record.

Richie, B. (1996). *Compelled to crime: The gender entrapment of battered black women*. New York: Routledge.

Richie, B. (2000). Exploring the links between violence against women and women's involvement in illegal activity. In B. Richie, K. Tsenin, & C. Widom (Eds.), *Research on women and girls in the criminal justice system* (pp. 1–13). Washington, DC: National Institute of Justice.

Roberts, S. (1971, June 13). Crime rate of women up sharply over men's. *New York Times*, pp. 1, 72.

Robinson, R. (1990). *Violations of girlhood: A qualitative study of female delinquents and children in need of services in Massachusetts*. Unpublished doctoral dissertation, Brandeis University, Waltham, MA.

Rogers, K. (1972, Winter). "For her own protection . . .": Conditions of incarceration for female juvenile offenders in the state of Connecticut. *Law and Society Review*, 223–246.

Roiphe, K. (1993). *The morning after*. Boston: Little, Brown.

Rossi, A. (1973). *The feminist papers: From Adams to Beauvoir*. New York: Columbia University Press.

Rowe, D. C., Vazsonyi, A. T., & Flannery, D. J. (1995). Sex differences in crime: Do means and within-sex variation have similar causes? *Journal of Research in Crime and Delinquency, 31*(1), 84–100.

Russell, D. E. (1986). *The secret trauma: Incest in the lives of girls and women*. New York: Basic Books.

Santiago, D. (1992, February 23). Random victims of vengeance show teen crime: Troubled girls, troubling violence. *Philadelphia Inquirer*, p. A1.

Schiraldi, V., Kuyper, S., & Hewitt, S. (1996). *Young African Americans and the criminal justice system in California: Five years later*. San Francisco: Center on Juvenile and Criminal Justice.

Schlossman, S., & Wallach, S. (1978). The crime of precocious sexuality: Female juvenile delinquency in the Progressive Era. *Harvard Educational Review, 48*, 65–94.

Schur, E. (1984). *Labeling women deviant*. New York: Random House.

Schwarcz, S. K., Bolan, G. A., Fullilove, M., McCright, J., Fullilove, R., Kohn, R., et al. (1992). Crack cocaine and the exchange of sex for money or drugs. *Sexually Transmitted Diseases, 19*, 7–13.

Schwartz, I. M. (1989). *(In)Justice for juveniles: Rethinking the best interests of the child*. Lexington, MA: Lexington Books.

Schwartz, I. M., Jackson-Beeck, M., & Anderson, R. (1984). The "hidden" system of juvenile control. *Crime and Delinquency*, *30*, 371–385.

Schwartz, I. M., & Orlando, F. (1991). *Programming for young women in the juvenile justice system*. Ann Arbor: University of Michigan, Center for the Study of Youth Policy.

Schwartz, I. M., Steketee, M., & Schneider, V. (1990). Federal juvenile justice policy and the incarceration of girls. *Crime and Delinquency*, *36*, 503–520.

Sentencing Project. (2002b). *Facts about prisons and prisoners*. Washington, DC: Author.

Sentencing Project. (2002a). *New prison population trends show slowing of growth but uncertain trends*. Washington, DC: Author.

Sewenely, A. (1993, January 6). Sex abuse charges rock women's prison. *Detroit News*, pp. B1, B7.

Shacklady-Smith, L. (1978). Sexist assumptions and female delinquency. In C. Smart & B. Smart (Eds.), *Women and social control* (pp. 74–86). London: Routledge & Kegan Paul.

Shaw, C. R. (1930). *The jack-roller*. Chicago: University of Chicago Press.

Shaw, C. R. (1938). *Brothers in crime*. Chicago: University of Chicago Press.

Shaw, C. R., & McKay, H. D. (1942). *Juvenile delinquency in urban areas*. Chicago: University of Chicago Press.

Shelden, R. (1981). Sex discrimination in the juvenile justice system: Memphis, Tennessee, 1900–1971. In M. Q. Warren (Ed.), *Comparing male and female offenders* (pp. 55–72). Beverly Hills, CA: Sage.

Shelden, R. G., & Horvath, J. (1986, February-March). *Processing offenders in a juvenile court: A comparison of males and females*. Paper presented at the annual meeting of the Western Society of Criminology, Newport Beach, CA.

Shelden, R. G., Snodgrass, T., & Snodgrass, P. (1993). Comparing gang and non-gang offenders: Some tentative findings. *The Gang Journal*, *1*, 73–85.

Sherman, F. (2002, August/September). Promoting justice in an unjust system: Part two. *Women, Girls & Criminal Justice, 3*, 65–80.

Shorter, A. D., Schaffner, L., Shick, S., & Frappier, N. S. (1996). *Out of sight, out of mind: The plight of girls in the San Francisco juvenile justice system*. San Francisco: Center for Juvenile and Criminal Justice.

Sickmund, M. (2000). *Offenders in juvenile court, 1997*. Washington, DC: Office of Juvenile Justice and Delinquency Prevention.

Siegal, N. (1995, October 4). Where the girls are. *San Francisco Bay Guardian*, pp. 19–20.

Simon, R. (1975). *Women and crime*. Lexington, MA: Lexington Books.

Simon, R. J., & Landis, J. (1991). *The crimes women commit, the punishments they receive*. Lexington, MA: Lexington Books.

Simons, R. L., Miller, M. G., & Aigner, S. M. (1980). Contemporary theories of deviance and female delinquency: An empirical test. *Journal of Research in Crime and Delinquency*, *17*, 42–57.

Sinclair, A. (1956). *The better half*. New York: Harper & Row.

Singer, L. R. (1973). Women and the correctional process. *American Criminal Law Review, 11*, 295–308.

Smart, C. (1976). *Women, crime and criminology: A feminist critique.* London: Routledge & Kegan Paul.

Smith, D. E. (1992). Whistling women: Reflections on rage and rationality. In W. K. Carroll, L. Christiansen-Ruffman, R. F. Currie, & D. Harrison (Eds.), *Fragile truths: 25 years of sociology and anthropology in Canada* (pp. 207–226). Ottawa, Canada: Carleton University Press.

Smith, D., & Paternoster, R. (1987). The gender gap in theories of deviance: Issues and evidence. *Journal of Research in Crime and Delinquency, 24*, 140–172.

Snell, T. L., & Morton, D. C. (1994). *Women in prison. (Special report).* Washington, DC: Bureau of Justice Statistics.

Sommers, I., & Baskin, D. (1992). Sex, race, age, and violent offending. *Violence and Its Victims, 7*(3), 191–201.

Sommers, I., & Baskin, D. (1993). The situational context of violent female offending. *Crime and Delinquency, 30*, 136–162.

Sommers, I., Baskin, D., & Fagan, J. (2000). *Workin hard for the money: The social and economic lives of women drug sellers.* Huntington, NY: NOVA.

Stahl, A. L. (2001). *Delinquency cases in juvenile courts, 1998.* Washington, DC: U.S. Department of Justice, Office of Juvenile Justice and Delinquency Prevention.

Stark, E., Flitcraft, A., Zuckerman, D., Grey, A., Robison, J., & Frazier, W. (1981). *Wife abuse in the medical setting: An introduction for health personnel.* Domestic Violence Monograph Services, No. 7. Rockville, MD: National Clearinghouse on Domestic Violence.

Steffensmeier, D. J. (1980). Sex differences in patterns of adult crime, 1965–1977. *Social Forces, 58*, 1080–1108.

Steffensmeier, D., & Allan, E. (1995). Criminal behavior: Gender and age. In J. Sheley (Ed.), *Criminology: A contemporary handbook* (pp. 83–114) Florence, KY: Wadsworth.

Steffensmeier, D. J., & Steffensmeier, R. H. (1980). Trends in female delinquency: An examination of arrest, juvenile court, self-report, and field data. *Criminology, 18*, 62–85.

Stein, B. (1996, July). Life in prison: Sexual abuse. *The Progressive*, pp. 23–24.

Straus, M. A., Gelles, R. J., & Steinmetz, S. (1980). *Behind closed doors: Violence in the American family.* Garden City, NY: Doubleday.

Streib, V. (2002). *Death penalties for female offenders.* Retrieved December 15, 2002, from www.law.onu.edu/faculty/streib/femdeath.htm

Studervant, S. P., & Stoltzfus, B. (1992). *Let the good times roll: Prostitution and the U.S. military in Asia.* New York: New Press.

Sutherland, E. (1978). Differential association. In B. Krisberg & J. Austin (Eds.), *Children of Ismael: Critical perspectives on juvenile justice* (pp. 128–131). Palo Alto, CA: Mayfield.

Sutherland, E., & Cressey, D. (1978). *Criminology* (10th ed.). Philadelphia: Lippincott.

Szerlag, H. (1996, February). Teen's death probed at YSI Iowa center. *Youth Today*, pp. 42, 48.

Tappan, P. (1947). *Delinquent girls in court*. New York: Columbia University Press.

Task Force on Juvenile Female Offenders. (1991). *Young women in Virginia's juvenile justice system: Where do they belong?* Richmond, VA: Department of Youth and Family Services.

Taylor, C. (1990). *Dangerous society*. East Lansing: Michigan State University Press.

Taylor, C. (1993). *Girls, gangs, women and drugs*. East Lansing: Michigan State University Press.

Teilmann, K. S., & Landry, P. H., Jr. (1981). Gender bias in juvenile justice. *Journal of Research in Crime and Delinquency, 18*, 47–80.

Thorne, B. (1993). *Gender play: Girls and boys in school*. New Brunswick, NJ: Rutgers University Press.

Thrasher, F. M. (1927). *The gang*. Chicago: University of Chicago Press.

Toby, J. (1957). Social disorganization and stake in conformity: Complementary factors in predatory behavior of hoodlums. *Journal of Criminal Law, Criminology and Police Science, 48*, 12–17.

Tracy, P. E., Wolfgang, M. E., & Figlio, R. M. (1985). *Delinquency in two birth cohorts*. Washington, DC: U.S. Department of Justice.

U.S. Department of Justice, Office of Justice Programs. (1989). *Children in custody, 1975–1985*. Washington, DC: Author.

U.S. House of Representatives. (1992). *Hearings on the reauthorization of the Juvenile Justice and Delinquency Prevention Act of 1974: Hearings before the Subcommittee on Human Resources of the Committee on Education and Labor*. 102nd Congress, Serial No. 102–125. Washington, DC: Government Printing Office.

U.S. House of Representatives, Subcommittee on Human Resources of the Committee on Education and Labor. (1980). *Juvenile justice amendments of 1980*. Washington, DC: Government Printing Office.

U.S. Statutes at Large. Ninety-Sixth Congress, 2nd sess. (1981). *Public Law 96–509—December 1981*. Washington, DC: Government Printing Office.

Valentine Foundation and Women's Way. (1990). *A conversation about girls*. (Pamphlet). Bryn Mawr, PA: Valentine Foundation.

Vedder, C. B., & Somerville, D. B. (1970). *The delinquent girl*. Springfield, IL: Charles C Thomas.

Vigil, D. (1995). Barrio gangs: Street life and identity in Southern California. In M. W. Klein, C. L. Maxson, & J. Miller (Eds.), *The modern gang reader* (pp. 125–131). Los Angeles: Roxbury.

Watson, T. (1992, November 16). Ga. indictments charge abuse of female inmates. *USA Today*, pp. A3.

Weis, J. G. (1976). Liberation and crime: The invention of the new female criminal. *Crime and Social Justice, 6*, 17.

Weithorn, L. A. (1988). Mental hospitalization of troublesome youth: An analysis of skyrocketing admission rates. *Stanford Law Review, 40*, 773–838.

Weller, R. (1996, July 11). Teens returned to parents after claiming abuse by camp counselors. *Associated Press*, pp. 1–2.

Wells-Parker, E., Pang, M. G., Anderson, B. J., McMillen, D. L., & Miller, D. I. (1991). Female DUI offenders. *Journal of Studies in Alcohol, 52*, 142–147.

White, J. W., & Kowalski, R. (1994). Deconstructing the myth of the nonaggressive woman. *Psychology of Women Quarterly, 18*, 487–508.

Widom, C. S. (1988). *Child abuse, neglect, and violent criminal behavior*. Unpublished manuscript.

Widom, C. S. (2000). Childhood victimization and the derailment of girls and women to the criminal justice system. In B. Richie, K. Tsenin, & C. Widom (Eds.), *Research on women and girls in the criminal justice system* (pp. 27–36). Washington, DC: National Institute of Justice.

Widom, C. S., & Kuhns, J. B. (1996). Childhood victimization and subsequent risk for promiscuity, prostitution, and teenage pregnancy: A prospective study. *American Journal of Public Health, 86*, 1607–1610.

Williams, T. (1992). *Crackhouse: Notes from the end of the line*. New York: Penguin.

Wilt, S., & Olson, S. (1996). Prevalence of domestic violence in the United States. *Journal of the American Medical Women's Association, 51*(3), 77–83.

Winick, C., & Kinsie, P. M. (1971). *The lively commerce: Prostitution in the United States*. New York: New American Library.

Wolf, N. (1993). *Fire with fire*. New York: Random House.

Yamamoto, E. (1979). The significance of local. *Social Process in Hawaii, 27*, 101–115.

Young, A. M., Boyd, C., & Hubbell, A. (2000). Prostitution, drug use, and coping with psychological distress. *Journal of Drug Issues, 30*, 789–800.

Youth Risk Behavior Surveillance System. (2002). *The 2001 Youth Risk Behavior Survey summary and results*. Atlanta, GA: Centers for Disease Control. Available online at: www.cdc.gov.

Yumori, W. C., & Loos, G. P. (1985). *The perceived service needs of pregnant and parenting teens and adults on the Waianae Coast* (Working Paper). Honolulu, HI: Kamehameha Schools/Bishop Estate.

Zane N., & Sasao, T. (1992). Research on drug abuse among Asian Pacific Americans. *Drugs and Society, 304*, 181–209.

Zatz, M. S. (1985). Los Cholos: Legal processing of Chicano gang members. *Social Problems, 33*, 13–30.

Zietz, D. (1981). *Women who embezzle or defraud: A study of convicted felons*. New York: Praeger.

INDEX

ABOUT THE AUTHORS

———•••———

Meda Chesney-Lind, Ph.D., is Professor of Women's Studies at the University of Hawaii at Manoa. She has served as Vice President of the American Society of Criminology and president of the Western Society of Criminology. Nationally recognized for her work on women and crime, her books include *Girls, Delinquency, and Juvenile Justice* (1992), which was awarded the American Society of Criminology's Michael J. Hindelang Award for the "outstanding contribution to criminology, 1992," *The Female Offender: Girls, Women, and Crime* (1997), *Female Gangs in America* (1999), and *Invisible Punishment* (2002). She is currently at work on an edited collection, *Girls, Women, and Crime,* in addition to this update of *The Female Offender.* She received the Bruce Smith, Sr., Award "for outstanding contributions to Criminal Justice" from the Academy of Criminal Justice Sciences in April 2001. She was named a fellow of the American Society of Criminology in 1996 and has also received the Herbert Block Award for service to the society and the profession from the American Society of Criminology. She has also received the Donald Cressey Award from the National Council on Crime and Delinquency for "outstanding contributions to the field of criminology," the Founders award of the Western Society of Criminology for "significant improvement of the quality of justice," and the University of Hawaii Board of Regent's Medal for "Excellence in Research."

In Hawaii, Chesney-Lind has served as Principal Investigator of a long-standing project on Hawaii's youth gang problem funded by the State of Hawaii Office of Youth Services. She has more recently also received funding to conduct research on the unique problems of girls' at risk of becoming delinquent from the Office of Juvenile Justice and Delinquency Prevention. Finally, she has also been tapped by the Hawaii Department of Public Safety to serve on an advisory panel on the problems of women in prison in Hawaii.

Lisa Pasko is a Sociology Ph.D. candidate at the University of Hawaii at Manoa. She received her master's degree in sociology from the University of Nevada, Reno, and has been involved in criminal justice research for more than 7 years. She is currently Project Coordinator for the University of Hawaii at Manoa Youth Gang Project. In 1999, she was Student Representative for the Western Society of Criminology, and in 2000 she received the American Society of Criminology, Critical Criminology Division, first place graduate paper award for her manuscript "Criminal Justice in the Mother Tongue: A Feminist Critique of Restorative Justice." Her publications include an examination of ethnic disparities in federal drug offense sentencing and an investigation of stripping in Hawaii. In addition to drug and sex work research, her interests include girls and the juvenile justice system, masculinities and crime, and gender theory.